Supervision in Psychodrama

Hannes Krall • Jutta Fürst
Pierre Fontaine (Eds.)

Supervision in Psychodrama

Experiential Learning in Psychotherapy and Training

Springer VS

RESEARCH

Hannes Krall
Klagenfurt, Austria

Pierre Fontaine
Louvain, Belgium

Jutta Fürst
Innsbruck, Austria

Published with the support of the Forschungsrat of Universitaet Klagenfurt with sponsoring provided by Privatstiftung Kaerntner Sparkasse.

ISBN 978-3-531-19678-7
ISBN 978-3-531-19679-4 (eBook)
DOI 10.1007/978-3-531-19679-4

The Deutsche Nationalbibliothek lists this publication in the Deutsche Nationalbibliografie; detailed bibliographic data are available in the Internet at http://dnb.d-nb.de.

Springer VS

Printed on acid-free paper

Springer VS is a brand of Springer DE. Springer DE is part of Springer Science+Business Media.
www.springer-vs.de

Contents

II. Relational Aspects between Supervisor, Supervisee and Client

III. Specific Methods in Supervision

IV. Training and Research in Supervision

Supervision in psychodrama – an introduction

Hannes Krall, Jutta Fürst & Pierre Fontaine

Supervision plays an important role in the process of learning and professional development in psychotherapy and training. Trainees and psychodrama practitioners reflect upon their experiences to see what has been achieved, what went wrong, what could be done in a different way. No matter which theory and methodological approach is applied, supervision seems to be essential for professional learning.

Many different approaches and definitions of supervision can be found in the literature. Bernard and Goodyear (2009) published a synthesis of literature on clinical supervision in general. Supervision is defined as

"*an intervention provided by a more senior member of the profession to a more junior member of that same profession. This relationship is evaluative and hierarchical, extends over time, and has the simultaneous purposes of enhancing the professional functioning of the more junior person(s): monitoring the quality of professional services offered to the clients that she, he or they see; and serving as gatekeeper for those who are to enter the particular profession*" (2009, p. 7).

This definition can serve as a major guideline. However, it has to be emphasised that supervision in psychodrama is an interactive learning process, rather than a one way directed intervention. It is hierarchical but – depending on the level of training – also an evolving relationship which eventually should lead to an equal relationship of experts. And supervision is evolving from directing how to work with the client or a group to an interactive reflection and investigation on a specific question or problem in order to improve psychodramatic work. In this way supervision can be seen as a co-creative effort to act as practitioner researchers to assure good quality but also to create specific and situated knowledge for further practice.

Supervision has a long tradition in different disciplines like social work, education, psychology, psychotherapy and medicine. Authors like Houtsma (1973, p. 10), Bernard and Goodyear (2009, p. 81) see the origin of supervision in Charity Organization in the United States. In the late nineteenth century they were overseeing their volunteers by agents of social work. These rewarded agents were charged with a double mission: to train these volunteers to work with the poor and the distressed and also to control the efficiency of work of these volunteers. These different aspects of supervision can still be found in current supervision. "The main goal of supervision is to foster the supervisee's professional development (a supportive and educational function) and to ensure client welfare" (Bernard & Goodyear, 2009, p. 12).

In the field of psychotherapy Freud (1909/1967) was probably the first supervisor when he reflected in a dialogue with a student the case of "Little Hans" who was the son of this student. Learning from practical experiences in psychotherapy was also important in other traditions. Rogers (1942) asked to register and replay the sessions. He advocated the same approach in supervision as in therapy. Bernard and Goodyear (2009, p. 83) report that "Supervision was a central and longstanding concern of Rogers".

Moreno was not explicitly referring to supervision or similar procedures. His focus was on the actors and what happened on stage in the here and now. Reflection or implicit 'supervision' takes place on stage or through action. He strongly believed that solutions can be found only by the involved persons themselves and not by experts who just have a distant view from outside. Moreno´s concept of learning seems to follow a kind of apprenticeship model. He encouraged trainees to be spontaneous, creative and to find their own way.

At present in psychodrama training – no matter whether it is applied in education, counseling or psychotherapy – supervision is an important step in the professional development of novices. Therefore, supervision is explicitly part of the training standards of the Federation of European Psychodrama Training Organisations (FEPTO).

During their training trainees are practicing psychotherapy in individual and group settings and they are accompanied by an ongoing reflective and investigative process of supervision with an experienced trainer in psychodrama. All practical and theoretical elements which were studied along the psychodrama training have to be put together to develop an individual working style which is based on common professional standards. High quality in supervision is a crucial part of the training, to support and to guide the trainee and to be a reliable point of reference in terms of practice and theory, technical and methodological standards, ethical issues etc.

Therefore, different formats of supervision are applied: supervision can take place in an individual or a group setting (co-therapeutic dyad, larger team, group). A special format is the peer supervision, where the trainees in a group are subsequently supervisor and supervisee. The common formats of supervision are:

- *Live-supervision within the training group:* A trainee is directing a group session with his or her peers who are in the role of clients. The supervisor is present and observing and reflecting the work of the trainee and sometimes intervening. It allows him or her to see the work of the trainee directly and to give feedback immediately. It is based on the presumption that a training group is a good enough simulation of the practical work with clients or at least comparable. This is a view which is not shared by all experts.

• *Indirect supervision of trainee's practice:* This is the most frequently used setting in supervision. The supervisor is not present at the session of the trainee with the client. In supervision trainees bring up their questions and challenges regarding their work, which they first present verbally, by roleplaying, by audio or video records. Usually trainees describe the difficulties and challenges they are facing in their work. Later on it can be a verbal reflection on the problem and the possible solutions or also a reflection by sociometric or psychodramatic methods. Supervision in this form is demanding a high level of disclosure.

• *Live-supervision or direct supervision in daily practice:* The trainee is working with a client or a group in in a clinic or in private practice. In direct supervision or live-supervision the supervisor is present at the session. The supervisor can be in the room or – what is very rarely practiced – behind a one-way-vision screen. In the room the supervisor can observe or take roles (consultant, co-director, etc.) Behind the mirror, the supervisor can observe alone or with a team (Anderson, 1987) while the session is conducted by the supervisee.

Supervision in psychodrama seems to be based on a variety of different practices, which have developed over time in different countries. Therefore, a need for sharing concepts and experiences has become evident.

In the last years Pierre Fontaine initiated workshops for trainers during the annual meeting of FEPTO in order to enhance professional sharing and exchange of theories and practices in supervision. Since then he pursued the idea to edit a book on supervision in psychodrama training to introduce and discuss different approaches.

Psychodrama trainers from different countries were invited to write about their way of doing supervision. A call for papers was announced at the FEPTO homepage, and some potential authors were invited to contribute with their knowledge and experiences.

The main purpose of this book is to share our ways of doing supervision in training in different countries and to stimulate fruitful discussions and further developments on good quality in supervision. The focus on topics and questions asked for in this invitation included aspects like the following:

• *Theory, philosophy, ethics of important aspects of supervision:* theoretical, conceptual, philosophical, ethical issues, which are important to consider. What is the basis, background or frame of our supervision? What are the roles and relationship between supervisors and supervisees?

• *Method, settings, techniques, interventions in supervision:* methodological aspects which are important or stimulating for our work. How are we doing it? What are the goals, the measures, the guidelines for supervision? What are special aspects like work on the relationship between supervisor and supervisees?

• *Content and processes of supervision:* interesting experiences, themes, cases, sessions, developments, conflicts, examples of good practice, 'magic moments' and pitfalls. What are practical approaches and helpful aspects in supervision?

• *Research and evaluation of supervision:* research, supervision or intervision for supervisors, quality assurance, evaluation. How do we know what works in supervision? How do we improve supervision?

• *Training of supervisors:* How is the training for supervisors done? Are there any assessment procedures? How many years of practice and self-reflection are required?

The contributions represented in the book show a variety of different theoretical concepts and practices in supervision. It is obvious that over time a rich basis of different concepts and practices has created a good basis for learning in psychodrama. Nevertheless, there are lots of challenges for future discussions and developments. Theoretical foundations, ethics, training and research on supervision can be considered as important challenges for further reflection.

As editors we are deeply grateful to all authors who have shared their approaches and experiences in supervision and thereby contributed to this book. We are also very grateful to Dorothy Langley and Peter Kohler for their assistance with English corrections. We also want to thank the Federation of European Psychodrama Training Organisations (FEPTO) and the University of Klagenfurt for their support in publishing this book.

We hope that the chapters in this book can provide an informative and encouraging resource, which stimulates further creative developments in theory, research and practice.

References

Anderson, T. (1987). The reflecting team. Dialogue and metadialogue in clinical work. *Family Process 26,* 415-428.

Bernard, J. M., & Goodyear, R. K. (1992, 2009, 4th ed.). *Fundamentals of clinical supervision.* Upper Saddle River (N. J.): Pearson Education.

Freud, S. (1909/1967). Analyse d'une phobie chez un petit garçon de 5 ans (Le petit Hans). *Cinq psychanalyses* (pp. 93-198). Paris: Presses Universitaires de France.

Houtsma, L. (1973). *Aspecten van supervisie.* Bloemendaal (NL): H. Nelissen.

I. Theoretical and Conceptual Frames

Theoretical and conceptual frames – an introduction

Jutta Fürst

When we view the various ways and forms of supervision in training it is like strolling through different beautiful gardens. When we experience pleasure in the variety being presented we can avoid questioning which one is the better. We can enjoy with less criticism. Looking at a certain arrangement we will consider copying it at our own place but maybe it becomes obvious that our garden has a different ground, another climate, another size and it will not transfer to our situation.

Using the metaphor of a garden, concepts and frames of supervision can be understood as the soil and the design of the garden. Different plants thrive in different situations and conditions. In the same way, supervision techniques which work well for some people may not be so effective for others.

Supervision in training is a learning process which involves reflecting on the practical work of trainees. The way in which it is done is based on different concepts and frames of reference. It is influenced by philosophy, anthropology, methodology and learning theories that are most valued by the supervisor and the training institute.

Some of the main questions being considered in anthropology are not only influencing the method but also the underlying philosophies behind supervision; questions such as "Who are we?", "How do we relate to other people and beings?", "How are we affected by them?" and "How free are we in our decision making?"

The theoretical concept of psychodrama is based on the encounter in the here and now, on action, on the potential for creativity and spontaneity in all beings, on the development in roles that exist from the very beginning, on the sociometrical structure of relationships and on the group as media for change (Buer, 1999). Psychodramatic supervision also draws on the same concept (Krall, 2008).

But as we can see in the following chapters we do not have only pure psychodrama supervision in the training processes. Psychoanalytic, systemic, cognitive behavioral and humanistic approaches have been integrated. They vary in their theory and techniques but also in their philosophical and anthropological concepts. Although Moreno tried to demarcate psychodrama from psychoanalysis permanently from his first psychodramatic steps in Vienna (Moreno, 1995;

Moreno, 1954) the developments over the years brought many fruitful crossovers of these two methods (Anzieu, 1984; Kellermann, 1980; Lebovici, 1971; Lemoin 1982; Widlöcher, 1974). It also led to other approaches which seem only on first sight incompatible.

Psychodrama and the systemic method were combined by Williams (1995), Farmer (1998), Lauterbach (2003), Varga von Kibed (2008) et al. Also the client-centered approach found its entrance into Psychodrama (Shearon, 1981; Macdonald, 1985) and the cognitive therapy enters into a partnership with psychodrama (Boury, Treadwell & Kumar, 2001; Baim 2007) as well.

The contributions in this section present a small selection of options for the ways in which supervision can be conceptualised and structured. This sample could be seen as an appetizer for the following sections and for the exchange of knowledge between supervisors in the field. It should also help to reflect the own style of supervising and the concept on which the supervision is maybe unconsciously based on.

Norbert Apter focuses in his chapter on the main attitudes of a Rogerian therapist, being congruent, empathic and understanding, with an emphasis on the effectiveness of feedback. Although giving feedback is a basic element of all training and supervision methods and something very familiar it reminds us how important proper feedback is. In an example the author is describing supervision of the practical work of the supervisees.

He created the method of "reverse enactment" in which the supervisee, who is bringing a case, is playing only the role of the client and never the role of the therapist. The supervisor trusts in the ability of the supervisees to find the answers by themselves by self-observation and introspection. The Rogerian style fits into the psychodramatic approach of supervision without any difficulties.

Maurizio Gasseau and Leandra Perrotta delineate "supervision in situ" a way of live supervision within a Jungian psychodrama training group. Psychoanalytic supervision in general oscillates between two conceptual extremes of the analysis of counter transference and an educational position. In the Jungian psychodrama the focus is furthermore laid on finding the collective dimension of the imaginations of a group as a key to understand what is going on (Barz, 1988; Scategni, 1994).

The depicted structure of supervision by Gasseau and Perrotta separates the didactic part of the supervision from the analytic part by having two different sessions. In the first part the supervisor is mainly an observer, analyser and teacher helping the trainee in difficult situations with advice and feeding back his impressions after the trainee has finished. In the second part of the process the supervisor becomes the therapist/conductor and the supervisee takes the role of the protagonist.

Agnes Dudler and Kersti Weiß present a supervision model which focuses on the interaction and interrelatedness "between inner and outer reality and between psychological, physical and sociodramatic roles" Supervision in this sense is a process of learning to adjust one's own method to the professional field a person is working in. The concept is close to a systemic view of relations. The psychodramatic method is used to feel the interaction by enacting, to find solutions for problems emerging out of this interrelatedness and to use the advantages of the system.

Giovanni Boria in contrast focuses on learning by an ongoing process of observing, planning, performing, analysing, getting feedback and again reflecting with a strong cognitive and behavioural component. The application of videotaping allows a careful observation and focuses more on behaviour and thinking than on feelings and intuition. The structure of the training and supervisory process with learning in systematic steps has a strong didactic impact and includes a permanent self-evaluation and evaluation by the supervisor and the trainees' peer group.

Beside the advantages that can be seen in all these methods there are disadvantages too. To be trustful, open and honest can be quite difficult for a supervisee when the supervisor is evaluating the work of the trainee at the same time. In the so called "live-supervision" groups it might also be stressful for a trainee to take the role of the director in a group of peers and to perform well with competitive colleagues who are in the role of clients.

Finally Pierre Fontaine describes and discusses in his chapter different arrangements of supervision in psychodrama training. This is done by going through the phases of the training: the therapeutic and experiential phase, the training group, the internship as an assistant, and the supervised practice as an independent practitioner. Fontaine also reflects on the elements and philosophy of this training.

References

Anzieu, D. (1984). *Analytisches Psychodrama mit Kindern und Jugendlichen.* Paderborn: Junfermann.

Barz, E. (1988). *Selbstbegegnung im Spiel. Einführung in das Psychodrama.* Zürich: Kreuz Verlag.

Baim, C. (2007). Are you a cognitive psychodramatist? Reflections on the links between cognitive therapy and psychodrama. *The British Journal of Psychodrama & Sociodrama*, 22, 23-31.

Boury, M., Treadwell, T., & Kumar, V. K. (2001). Integrating psychodrama and cognitive therapy: an exploratory study. *The International Journal of Action Methods: Psychodrama, Skill Training, & Role Playing*, 54, 13-37.

Buer, F. (1999). *Morenos therapeutische Philosophie.* Opladen: Leske+Budrich.

Farmer, C. (1998). *Psychodrama und systemische Therapie.* Stuttgart: Klett Cotta.

Kellermann, P. F. (1980). Übertragung, Gegenübertragung und Tele - eine Studie der therapeutischen Beziehung in Psychoanalyse und Psychodrama. *Gruppenpsychotherapie und Gruppendynamik.* 15, 3/4, 188-205.

Krall, H. (2008). Psychodrama und Soziometrie in Supervision und Coaching.Anknüpfungspunkte in der qualitativen Sozialforschung. In H. Krall, E. Mikula, & W. Jansche W. (Eds.), *Supervision und Coaching* (pp. 252-268). Wiesbaden: VS Verlag für Sozialwissenschaften.

Lauterbach, M. (2003). Rollentausch und systemische Therapie. *Zeitschrift für Psychodrama und Soziometrie*, 1, 80-90.

Lebovici, S. (1971). Eine Verbindung von Psychodrama und Gruppenpsychotherapie. In De Schill St. (Ed), *Psychoanalytische Therapie in Gruppen* (pp. 313-339). Stuttgart: Klett.

Lemoin, G. (1982). Zu einer psychoanalytischen Theorie des Psychodramas. In H. Petzold (Ed.), *Dramatische Therapie* (pp. 127-147). Stuttgart: Hippokrates.

Macdonald, M. (1985). Das Psychodrama erforscht eine private Welt. Ein Praxisvergleich der Behandlungsansätze von Moreno und Rogers (1947). *Integrative Therapie*, 11 (1), 4-25.

Moreno, J. L. (1974). *Die Grundlagen der Soziometrie.* Opladen: Westdeutscher Verlag.

Moreno, J. L. (1995). *Auszüge aus der Biographie.* (Ed. Moreno J. D.) Köln: InSzenario.

Scategni,W. (1994). *Das Psychodrama. Zwischen alltäglicher und archetypischer Erfahrungswelt.* Solothurn, Düsseldorf: Walter.

Shearon, E. M. (1981). Ein Vergleich zwischen Rogers Selbst -Theorie und Morenos Spontaneitäts - Theorie. *Gruppendynamik*, 12 (3), 236-256.

Varga von Kibed, M. (2008). Systemische Strukturaufstellung als erlernbare Sprache. In H. Krall, E. Mikula, E., & Jansche W. (Eds.), *Supervision und Coaching* (pp. 97-108). Wiesbaden: VS Verlag für Sozialwissenschaften.

Widlöcher, D. (1974). *Das Psychodrama mit Jugendlichen.* Freiburg: Olten.

Williams, A. (1995). *Visual & Active Supervision.* New York, London: W.W. Norton & Company.

Humanistic processing: the supervisor's role through reverse enactment

Norbert Apter

The article will discuss the basic principles of supervision and thus of processing in humanistic psychodrama. On one hand supervision represents the necessary professional and personal safety (for the psychodramatist as well as the protagonist), on the other hand it offers assurance for an ever evolving therapy. Supervision in humanistic psychodrama centres on processing, it is therefore important that the supervisor masters the following challenges: one's role of facilitator (way of being), processing's focalisations (relations – method – clinic) and functions (for the trainee, the group, the supervisor-trainer), as well as both operational pillars of the framework (i.e. the rules of constructive feedback, and the general structuring of a processing: in this chapter we will also have a detailed look at the unfolding of reversed enactment processing.

1. The importance of supervision

"A professional therapist who remains without supervision for any length of time is in danger, indeed dangerous". That was the unsettling statement I heard from one of my teachers at Harvard while I was a student. At first it came as a surprise to me: I was at a state in my life when I still believed that once I had finished my studies, once my psychotherapy – person centered approach and psychodrama – and my specialization would be behind me, there might perhaps still be one or the other training or refresher course to renew my inspiration and that would be it.

It came as a big surprise to me when, in the course of my various specialised training programmes in psychotherapy – and through experience gained in my own practice – I discovered that supervision was a true benefit. In the course of my sessions in supervision I became aware of the diverse advantages it generates. Depending on my inclination, my needs, my moments, depending on the questions I had and brought to the meeting, I was able to share, clarify and untangle situations which previously had seemed difficult. I was able to overcome a lot of professional stress which might have become hazardous not only to me (worries, fatigue, burn out, etc.), but also to my patients (my becoming unavailable, a poor listener, irritable; the process might even have stalled all together). I came away from supervision with new directions to pursue a deeper investigation or with a renewed perspective. In other words, I used supervision to my advantage, and to the advantage of my professional environment, I was able to replenish and find renewed sources of energy and inspiration in order to become constructive (again).

The same applied to my colleagues, be they psychotherapists, physicians, social workers, teachers, educators, school principles, etc. Each and everyone was able to benefit from supervision, to resolve some deadlock, extend his/her scope of competence and efficiency, and increase the joy and gratification generated by his/her work. That is the reason why I have been involved as a supervisor for many years with professionals, individuals and teams, inside and outside of institutions, in Geneva and in the surrounding region.

I gradually realised that accepting supervision was a professional choice which from a cultural point of view is not always easy to make. As professionals, we are supposed to be able to recognise others, listen to them, hear them in a safe environment and help them. And even though that does not eliminate our own need for support and assistance, many of us have a hard time expressing that and accepting supervision. For taking that step, i.e. being supervised, implies: to simply recognise without any shame that we are not perfect and that that is ok; to acknowledge that we cannot always manage to do everything all alone; to choose to take a moment, space and means to share, clarify, expand, thus to learn from our imperfections and our resources.

Supervision in psychotherapy is indeed that necessary moment it takes to step back and give oneself some distance while being supported. Supervision promotes both a developmental process of our competences and a checking or controlling process of our professional practice. Through my experience both as a supervisee and a supervisor I've become strongly aware that supervision allows us…

- to question our practice;
- to find our theoretical and clinical marks for therapy again;
- to understand the underlying links in the relationship with our patients and to be able to move beyond those links, transferences or counter-transferences which sometimes hamper the smooth unfolding of the therapeutic process;
- to consider our own limits, thus being able to manage our stress and avoid emotional or organizational overload (or even burn-out);
- to be able to commit professionally, depending on the situation, to the quest for constructive and effective options, and therefore to the renewed dynamics of our practice.

Thus we will be able to both make our practice safe and assure the client, the institution and ourselves, of constantly evolving professional competences. So my teacher was right…

The following chapter explains the basic principles of processing in humanistic psychodrama[1] – after warm-up, enactment and sharing – the fourth phase of a session in the training process.

2. Humanistic processing

To my great surprise, processing, which is the fourth phase in psychodrama and which is so frequently used to provide supervision in training programmes throughout the world, is not mentioned very often in the relevant literature. James Sacks'[2] extensive bibliography contains some references, i.e. in Portuguese (e.g. Aguiar & Tassinari, 1994; Campedelli & Wolff, 1980; Monteiro & Brito, 2000; Naffah Neto et al., 1981) and in English (e.g. Goldman & Morrison, 1984; Kellermann, 1992; Williams, 1989). But very little seems to be written on the subject considering the vast amount of publications on psychodrama in general.

Processing is a moment to step back and gain some distance in order to analyse the various processes in a psychodrama session, both in order to verify one's practice and to develop further professional skills; it takes place in a supportive environment. As seen above, processing is a privileged mode for psychodrama supervision, both for supervision of the learning process and for supervision of professional practice. It is therefore important to master the challenges of processing in order to assure its smooth unfolding, whatever the focus at a given moment may be. This is even more important in a group setting (which is the setting we have chosen to develop below).

In humanistic psychodrama the role of the supervisor/trainer is primordial. On one hand he is the guarantor of all three intrinsic functions of processing: stepping back and gaining some distance whilst being supported; verifying or checking of professional practice; further developing of professional skills. On the other hand, the supervisor simultaneously implements five components: (1) taking over the role of the facilitator; (2) generating humanistic processing's essential triple focus (see 2.2); (3) offering the benefit of various processing

[1] Some consider psychodrama as a humanistic method per se. Some others do not. Therefore we have chosen to call 'humanistic psychodrama' that type of psychodrama which integrates both J. L. Moreno's method and Carl R. Rogers Person Centered Approach (Rogers is considered to be the founder of the humanistic approach).

[2] available under http://www.pdbib.org

functions; (4) guaranteeing constructive feedback; (5) effectively structuring the unfolding of processing.

Each one of these components is important and is made up of various elements which the supervisor will have to consider in supporting a trainee in humanistic psychodrama. I will develop these five aspects of the supervisor's role in the following chapter.

2.1 The role of the facilitator

In psychodrama, major importance is given to a secure relational climate. "Moreno's psychodrama [...] takes place in a setting accepting a human being's complexities and its interiorised roles. There is no judgment, analyses or interpretation" (Apter, 2003). Not during warm ups nor during enactment or sharing. This is what affords the protagonist fundamental safety and security. It remains compelling for processing, even if the latter part not in a facilitating climate. Psychodrama literature hardly elaborates on how to implement such a climate. Hence humanistic psychodrama refers to Carl Rogers (Rogers, 1968) for its relational dimension. Thus the fundamental role of the psychodramatist, and specifically of the supervisor in psychodrama, becomes clearly that of facilitator. For each psychodrama session, including processing, the facilitator looks to create a climate defined by three attitudes which together are necessary and sufficient:

• *unconditional acceptance of the human being* (also called *unconditional positive regard*). "Unrestricted acknowledgement encourages self-acknowledgement, self-acceptance and self-esteem. [...] Its aim is mutual recognition between two persons, not knowledge of the other person" (Schmid, 2001, p. 51). This welcoming of each other and acknowledgement of each other's being, whatever one may have done, promotes the right to make mistakes during the learning process and promotes the learning process itself (Rogers, 1972).

• *empathy,* the delicate and genuine attempt to understand the other person (Bohart & Greenberg, 1997; Rogers, 1975). In going towards each member of the group in an exploratory fashion, the supervisor/facilitator enables any difference or similarity in intelligence to surface gently, allowing each member to feel at ease in expressing himself, as well as enhancing the general understanding of the analysed process (Rogers, 1972).

• *congruence* is a specific type of authenticity, when a person expresses what he/she experiences and what he/she has drawn from it, owning it. This particular way of expressing and revealing oneself to another person tends to generate genuine contact, from one complete human being to another complete human being, and can lead to relational depth (Mearns & Thorne, 1999). When congruence coexists with acceptance and empathy, it can promote that type of dialogue and enhance a genuine encounter between I and Thou, so vital to Martin Buber and J. L. Moreno (Buber, 1969; Moreno, 1964).

Facilitating a relational climate is thus not 'a way of doing', but rather 'a way of being', even if it can be learned, which promotes safety and trust (Apter, 1987). Once a facilitated climate of safety and trust rules each person's innate tendency towards self-actualization[3] which can unfold on a developmentally favorable terrain (Rogers & Kinget, 1962); or – to use Fine's description of Moreno's theory – "the optimal spontaneity and adaptability factor" is activated (Fine, 1979). Thus each group member can contribute to the development of processing, individually or collectively, drawing on his/her freedom to analyse and his/her own resources; together they all benefit from it.

2.2 Triple focusing in humanistic processing

Jinnie Jefferies in one of her writings (in Karp, 1998) emphasises that analysing the process and its definition will depend on their context. She underlines some basic elements that it might be good to 'process': one might choose to analyse the group process, the protagonist's or the psychodramatist's process or even the trainer's. We all agree that focusing during processing may vary with the moment in time or the needs arising. Focusing in humanistic processing is essentially based on three major dimensions in psychodrama:

The relationship between the psychodramatist, the protagonist and the group: The relationship and its entailing inter-influences are at the centre of psychodrama (Moreno, 1937). As underlined by some humanistic psychologists (Maslow, 1972; Rogers, 1951), the existing relational climate during a session is of primordial importance in the patient's development. One of the characteristic features of humanistic psychodrama is therefore the special attention given to the psychodramatist's way of being and the expected climate of trust and confidence, safety and creative spontaneity that stem from it.

The use of psychodrama: A psychodramatist-in-training, thriving to master different aspects of this method, requires feedback on the way in which he/she took on the various roles of a psychodramatist, or on his capacity in situating an

[3] "Any human being, as valid as anyone else, will at any time consider internal and external circumstances and choose what is best for him - or least detrimental" (Apter, 2003). That is a basic principle in person centered psychotherapy, as well as in psychodrama's resource orientation developed by J. L. Moreno.

action in the here-and-now, or the way in which he chose and implemented the tools available in psychodrama, his use of effective techniques for better or for worse, etc.

Theoretical and clinical aspects: In psychotherapy, supervision through psychodrama can of course not occur without analysing diagnostic and prognostic elements in the development of a therapy plan, nor without thorough reflection on the link between theory and clinical observation leading to an accurate evaluation of therapy. This is an important part in processing. In humanistic processing, triple focusing is not only a formative goal, but also a prerequisite.

2.3 The various processing functions

During training, processing serves each trainee's learning (not only the one's directing the session). Processing helps in

- identifying significant elements in a session from a relational point of view, from a clinical one as well as from a methodological and technical one;
- analysing the above elements and their underlying dynamics;
- verifying and evaluating their adequacy in a given situation;
- linking together observed practice, theory and clinic;
- extending perspectives through developing possible options.

The supervised person maximises his/her learning process when the above processing functions have become active. I call these functions 'multi-individual functions'. Individual, for each trainee will use processing in his or her own way and multi, for all trainees, all group members will benefit from them.

Specific processing functions for the protagonist: The protagonist, who is also a learner in a training group, is no longer the center of attention or interest of the group. In fact, it is vital to see to it that this be not the case. For processing must not be seen as a continuation of the protagonist's therapeutic process. It would be most destabilizing for the protagonist to go over any interpretation of his life experience during a session. This would be totally counterproductive, as interpretation in psychodrama lies within the very action (Leutz, 1985). Thus there are two additional sine qua non functions of processing for the protagonist 'decentering' of his own process, and fully endorsing his role as a trainee.

Basic processing functions for the group: Processing is of key importance in the group's development and learning, its main use is to further the learning community through its own multiple intelligence.

Humanistic psychodrama promotes 'working together'. Processing supplements this basic function of the group's learning process. It develops interpersonal learning, facilitated by the trainer. Through processing, each trainee can put forward any question that has arisen and look for an answer, thus giving both himself and the group the possibility of 'gestalt' in the learning process. Through observation and questioning, each group member can call upon his manifold intelligence, or as Gardner (1996) would put it: his multiple intelligence.

In contributing their different styles of multiple intelligence, the group members allow the group to benefit from both individual and collective intelligences. Differences and similarities are being offset or supplement each other; each member contributes to the group's learning in many different ways. Each member's unique perspective (Apter, 1996) will contribute to a whole which is different from and 'superior' to the sum total of its parts. The learning group, this collaborative forum linked by common interest, develops its own body of knowledge thanks to its collective intelligence; the overall learning community stands to benefit from it.

Processing functions of the supervisor-trainer: This moment of multiple exchange enables the supervisor-trainer to:

• assess the supervised student's level as well as the level of the training group, and evaluate future steps required for the effective outcome of the training programme.

• acknowledge and validate the contributions of each individual member as well as of the group as a whole. On one hand, this strengthens the trainees' self-esteem and their hope to progress, on the other hand, it underlines the feeling of actively belonging to the group and contributing to its collective intelligence.

• summarise the lessons to be drawn and open gateways to the future: the supervisor determines the key learning elements and identifies those which require anchoring, extension or development during the subsequent module; he recognises the way ahead.

Processing thus has multiple functions which each participant and the whole group can put to good use. To allow for that to happen, it takes not only triple focusing and a facilitating relational climate, but furthermore, feedback must occur in a constructive manner.

2.4. The rules of constructive feedback

Processing, i.e. elaboration of the process of a psychodrama session is strongly based on quality and efficiency of the feedback. The role of the supervisor is to protect on one hand the feeling of safety and security and on the other hand the trusting and constructive climate established within the learning group.

Feedback is a mean of communication by which a person expresses in a nurturing way the image gained of one or several actions (behaviour, speech) contributed by another person. The very reason for feedback is a constructive one: basically it is used to nurture not only the person it is addressed to, but also the relationship, in the spirit of a co-constructive partnership.

To the extent that feedback can either facilitate or complicate not only relationships within the group, but also common tasks, trust, coherence, etc. it is vital that each participant prepare his/her feedback, i.e. (1) be aware of his intentions and finality, (2) identify his internal process, and (3) pay attention to existing sensitivities.

Once these elements have been considered, one then has to assess gradually how to best get the intended content across, whilst considering the existing relationship. The basic axiom underlying constructive feedback can be expressed in a simple sentence:

If feedback is received, it might be useful, therefore feedback needs to be conceived and expressed in such a way as to be receivable[4]. In order to increase receivability and reception of feedback, one must pay attention to the 'way of being' as well as to the 'way of expressing'.

The 'way of being' is based on the above-mentioned three attitudes (unconditional acceptance, empathy, congruence) which the supervisor must take great care in implementing. Sometimes students have difficulty experiencing all three attitudes together and keeping them active during their dialogue. They might feel empathy and acceptance and not dare express themselves clearly, or express themselves without paying attention to the other person's feelings, or alternatively pass quick judgment which might confuse the ongoing psychodramatist. It is therefore important to hint at 'ways of doing' during this fundamental moment of feedback. There are a few basic rules to observe in the 'way of being':

- Addressing the person as a unique person
- Addressing the person from the self
- Focusing on the person's behaviour, not on his/her being
- Describing, expressing practical observations without any judgment or interpretation
- Presenting a wide perspective, considering several elements, sometimes concurring, sometimes differing
- Remaining realistic, avoiding all generalization and exaggerations and generalizations; giving preference to precision (when? where? what?) and descriptive explanation;
- Acknowledging strong points and possible difficulties;

[4] If feedback is not receivable or not received, it will hardly be of use to the addressee.

- Keeping empathy's emotional distance, using our emotional intelligence (Goleman, 1995) and paying attention to the other person's emotional intelligence as well;
- Offering sufficient space for retro-action.[5]

Such personalised feedback might become receivable and might thus be received usable and might thus be used.

2.5 Structuring of processing

Explaining and implementing the above rules of feedback will however not be sufficient to preclude any possible glitches during supervision. Structuring of the processing therefore remains fundamental to promote a safe learning environment.

During direct supervision of an enacted situation in psychodrama, with one trainee as the director and another trainee as the protagonist, two people will expose themselves:

- *the protagonist, through a sometimes very complex issue:* The protagonist will have bared his inner/outer reality and let the group see it. Closing of the 2nd phase (enactment) and the 3rd phase (sharing) in psychodrama will have created a safe environment and will have enabled him to quietly grasp the outcome of his psychodrama. It is essential, and this must be stressed again, that during processing, his own psychological problems be not addressed, for this would be most unsettling.
- *the psychodramatist-in-training, in his specific way of being and doing:* When directing a psychodrama session in front of his fellow trainees, the student reveals his professional skills as well as the limits of his competences. Whatever the perceived quality of secure support he is given or whatever his openness and his wish for constructive conclusions, a psychodramatist-in-training who bared himself throughout his work in front of his peers and his supervisor may feel vulnerable. Therefore both of them, protagonist and psychodramatist, need to

[5] Giving feedback, showing the other person the image of himself that he/she prompts within us invariably implies revealing to him what kind of mirror we are. We all know what mirrors do. Their reflection is not always pleasing: either the image is too small, too big, too crude or it is slightly deformed. That is typical of mirrors. It is therefore not surprising to see the person addressed by the feedback wishing to speak up, to comment, to adjust, to refine, to clarify, to explain. Such retro-action is necessary after feedback. It is especially important in a group setting, as it avoids for a trainee to find himself powerless, full with all that feedback, including sometimes a bitter aftertaste. Giving the trainee the possibility to express his feelings in reaction to the feedback allows for completion of the interaction.

feel safe in their environment. It is reassuring for them to be able to trust the supervisor's methodology.

Processing typically unfolds in six steps in the following order: (1) self-feedback by the psychodramatist-in-training; (2) feedback from the protagonist; (3) feedback from members of the training group; (4) feedback from the supervisor; (5) retro-actions from the psychodramatist-in training; (6) perspective and closure.

The psychodramatist-in-training is the first one to assess his work, before anybody else tells him what they feel or think. In a way it is like recreating his/her professional reality: when the session is over and the patient has left, the psychodramatist proceeds to his self-evaluation – alone. The second feedback, the most significant one, is by the protagonist, in order to avoid any major influence from other participants on his own feedback, which somehow is the most important one. The third feedback comes from the peer group – this is when collective intelligence can become very potent, even before the supervisor gives his feedback to the psychodramatist-in-training. Afterwards, the latter can speak, he can clarify some feelings and draw some first conclusions. Before bringing the process to closure, the supervisor then contributes a meaningful perspective.

Processing, as mentioned earlier, is but one moment in supervision. In presenting the way in which supervision unfolds, some key details in the processing will appear and the multi-dimensional role of the supervisor throughout the various phases of supervision will be clarified.

The multi-dimensional role of the supervisor in humanistic psychodrama:

Carrying out all intrinsic functions in supervision: stepping back to gain perspective; checking against practice; developing the trainees' skills… by implementing 5 components simultaneously:

I. *Facilitating a positive relational climate:* Trusting the innate tendency towards self-actualization and promoting (a) congruence, (b) unconditional acceptance of each one, and (c) empathy.

II. *Generating humanistic processing's triple focus on:*
- the relationship between the psychodramatist, the protagonist and the group
- the use of the psychodramatic method and its techniques
- theoretical and clinical aspects

III. *Exploring processing's multiple functions:*
Benefits for the group: promoting a learning community
Benefits for the protagonist: facilitating his renewed role as a learner
Benefits for each trainee: identifying significant elements in a session, from a relational, methodological, technical as well as theoretical and clinical point of view; analysing the above elements and their dynamics; verifying and assessing their suitability in a given situation; linking observation, theory and clinical observation; extending perspectives.

Benefits for the supervisor: assessing general level and future steps required for further progress; recognise and value each member of the group and his/her contribution; summarise lessons learned, and indicate bridges to future learning.

IV. *Guaranteeing constructive feedback by implementing the rules:*
- Addressing the person as a unique human being
- Addressing him/her from one's inner self
- Focusing on a person's behaviour, not his being
- Describing, expressing practical observations
- Presenting a wide perspective
- Remaining realistic and avoiding all generalization
- Recognizing possible support elements as well as potential difficulties
- Maintaining empathy's emotional distance.
- Providing space for retro-action

V. *Ensuring the efficiency of processing in 6 steps through:*
(1) self-feedback from the psychodramatist; (2) feedback from the protagonist; (3) feedback from the members of the training group; (4) feedback from the supervisor; (5) retro-actions from the psychodramatist-in-training (6) perspective and closure: (a) draft synthesis/closure; (b) clarifying the results of reverse enactment; (c) final step back – closing.

Table 1: The multi-dimensional role of the supervisor in psychodrama

2.6. Unfolding of supervision through 'reverse enactment'

I would like to describe the unfolding of a specific type of group supervision, which is that of 'reverse enactment'. I call 'reverse enactment' an enactment throughout which the protagonist only plays the role of someone else (here his patient): at no point in the action phase of the psychodrama does the protagonist play his own role. It is actually my favourite supervision mode through enactment, for the lessons to be learned from it are numerous, substantial and fruitful.

The humanistic methodology, which is warranted by the supervisor, is based on each one of the theoretical steps identified earlier in this chapter leading to his subsequent multidimensional role, even if his role as facilitator allows him to adopt a non-intrusive attitude, sometimes even an outright discreet one. The construct of reverse action supervision is the following:

Identifying the problem, question or aim: The trainee brings one or several questions to the session: he explains to the group what it is he/she is bringing, the question on his mind on that particular day regarding one or several of his clients/patients (or even colleagues). The supervisor together with the trainee attempts to identify what the learner expects from his supervision: his aim, which needs to be realistic, reachable, verifiable and measurable.

John, a first year trainee asks for supervision in the monthly supervision group:
John: Clarissa has been a patient of mine for six months now. She suffers from depression. My problem is that she only comes on and off to the planned sessions, and I start being more and more angry at her.
Supervisor: What would you like to get from this supervision about it?
John: I would like to find a constructive way of confronting her.

Establishing a strategy: Together they will determine whether it is appropriate for the trainee to display his client/patient and to take on the role of the latter. If this appears to be beneficial to the learner, supervision through reverse enactment is offered. From that point onwards, the choice as to which of his fellow trainees will be the patient's psychodramatist is made by the trainee, with his agreement. The selected trainee will of course also need to agree (or the selection procedure will continue).

Implementing reverse action psychodrama: Hence all three phases of a psychodrama occur (warm up, acting, sharing) in which two trainees having a specific role:

• the protagonist: the trainee (John) who will attempt to solve the problem he/she brought to supervision and who will now proceed to a role reversal in becoming his own client (Clarissa), thus becoming the protagonist of the present psychodrama;

• the psychodramatist: the trainee (Paul) who will support the client (Clarissa) as perceived and represented by his fellow trainee (John) in moving forward in his own process.

Paul tells to Clarissa (played by John) that he has noticed how often she has been absent. Dialogue goes on for two minutes and suddenly Clarissa states: "I hate constraints, I've always been like this! When I was eighteen, my parents wanted me to be at home at least every Sunday for lunch, I hated it!"

All other trainees are available for further auxiliary roles: the psychodramatist may or may not call upon other group members (depending on whether he chooses to proceed to a group psychodrama or individual psychodrama). Sylvia and Rob are chosen to play the parents and the psychodrama goes on. The supervisor is available to the psychodramatist, should the latter need to rely on the supervisor.

Ensuring efficiency in processing: Once all three steps of psychodrama are concluded, 'the psychodramatist' (Paul), 'the protagonist' (John) having accomplished the role of his client (Clarissa) and all other group members revert to being trainees. The group now focuses on the trainee who was the 'psychodramatist' on his relations with the protagonist and with the group, on his use of pychodramatic method, and on the theoretical and clinical aspects at play in psychodrama. Thus processing can start[6], supported and facilitated by the supervisor.

Self-feedback: The trainee who directed a vignette or a session in psychodrama (Paul) is encouraged to express

- his experience as a psychodramatist (aim: awareness and expression of experience)
- his appreciation of his own work (aim: conservation and development);
- any difficulty or shortcoming he might have experienced (aim: correcting and/or repairing);
- possible options he could have chosen (aim: extension of possibilities);
- skills and competencies to be further developed (aim: development).

Feedback from the protagonist: It is important for the psychodramatist (Paul) to hear first of all his protagonist's feedback (Clarissa played by John), for in his real practice the protagonist will be his only partner; beyond the psychodramatist's experience (thoughts, feelings, etc.), only the patient knows and feels what seems right for her. This is not being questioned, for in humanistic psychotherapy the patient is the holder of his own truth.

In case there is a distortion between the psychodramatist's experience and that of his client, it might be a good time to look into the origin of such distortion – which in turn can lead to many insights, questions and assumptions as to the way of doing or being, or as to counter transference of the psychodramatist.

In order not to return to the client's role (Clarissa), the former protagonist (John), who previously played the role of his client and then quit the role, offers feedback on behalf of his client (Clarissa as represented and experienced by the

[6] In some particularly hazardous situations recalling the rules of feedback.

latter). The trainee speaks of his client as “he/she”. The psychodramatist-in-training remains silent for the time being.

Feedback from the members of the training group: Within the group, each member offers feedback to the psychodramatist-in-training (Paul). The latter remains silent, makes no comment, accepts possible questions but makes no response. Rob: “At the very beginning, I needed more clues about how to play the father, but little by little, the many role reversals helped me getting into the role.” The supervisor sees to it that all feedback remain constructive. If necessary, he rephrases, clarifies or reframes some of the comments.

Feedback from the supervisor: The supervisor in turn shares his analyses of the process. From the trainers I have collaborated with, I’ve been able to identify various types of feedback:

- feedback on the fly: quick feedback from memory about the key elements of a session and respectful analyses thereof;
- written feedback: the trainer carefully notes his observations, comments and questions throughout the session and expresses the essence thereof to the psychodramatist-in-training before handing him the notes.
- feedback-in-action: significant moments are being recreated and commented as a vignette on stage, sometimes they are amended by some proposals or options. Neither the protagonist nor the psychodramatist-in-training must be amongst the selected actors, for it is vital to avoid that the former start his psychodrama all over (or parts thereof) and to make sure that the latter gain the necessary perspective in this particular learning phase.

  The most original feedback that I received as a trainee came from Ken Sprague: on his sketch book he used to depict scenes on which he would base his feedback.

Retro-actions from the psychodramatist-in-training: So far the recipient of all feedback (Paul) has been silent, now he expresses his impressions regarding feedback. He/she is granted sufficient time to express his feelings, thoughts, emotions, etc. and to clarify them together with the group if need be. It is very important in my view to pay attention to the feelings of the psychodramatist-in-training at this stage, i.e. his pleasure in seeing some of his skills acknowledged, frustration (sometimes shame) in having some of his shortcomings identified, fear of ‘not making it’, sometimes even irritation or anger at being misunderstood. Granting the psychodramatist-in-training particular attention facilitates – if need be – his emotional process and thus his exploration thereof.

This is also the time when the psychodramatist-in-training is encouraged to elaborate what he/she thinks he can stand on and wishes to maintain (conservation); what he wishes to improve based on feedback (development); what he wishes to be particularly attentive to in order to develop a lacking skill (construction).

Perspective and closure: Draft synthesis/closure: When closing this 'central loop', the supervisor returns to various elements which have come up during the ongoing process, he formulates some conclusions reached by the psychodramatist-in-training, underlines the lessons learned and indicates some elements of progression.

Clarifying the result of reverse enactment: Now comes the time to close the 'first loop': the supervisor addresses the trainee (John) who had initially requested his supervision in a specific situation and who subsequently became the protagonist, (Clarissa – through role reversal): "At the beginning of this process your aim was... Where do you stand now? What do you conclude from this?"

The trainee (John) takes some time to assess the result of the process he has undergone, based on the outcome of his experience in the reversed role and the contributions of collective intelligence through the processing mechanism. John: "My anger towards Clarissa is significantly resembling her father's. Exploring with her what her absences are about would be more constructive and potentially efficient than demanding her regular presence!

He is invited to ponder relational dynamics involved, clinical indicators generated in the process, as well as the options he/she can now consider from a methodological and technical point of view. On each one of these focal points, the trainee expresses what appears to be

- options and questions to pursue in his practice,
- progress he believes he made,
- as well as the limitations of the process,
- and any irrelevant or pending aspects.

Gaining perspective and closure: This is where the session ends and the trainer-supervisor offers the group an ultimate possibility to take a step back and reflect in doing the following:

- anchoring the constructive features of the relations produced in the group during the process (climate);
- specifically recognising and valuing the learning community in its way of expressing itself, in linking practice, theory and clinic (collective and relational intelligence);
- drawing lessons to be learned from the session (gains through common experience);
- underlining the prospect of practical applications (possible practical applications);
- emphasising future learning steps (foreseen progression).

This last stage of gaining perspective is like the cherry on the cake for each member of the group. Every one of them individually and collectively can now anchor the awareness of insights and the difficulties met along the way, but it also enables the lessons drawn from the session to be directly applied in the trainees' professional practice whilst identifying remaining skills to be developed.

3. Conclusion

As seen in this chapter, supervision is a necessary and inherent element of a psychodramatist's training. Be it during a training module associating theory, self-experience and directing experience (processing of learning), or in the supervision of a professional practice (processing of practice), a trainee will have the opportunity to direct a session including all three typical phases in a psychodrama (warm up, action, sharing) with another group member as protagonist. Processing is added as a fourth phase, centred on the psychodramatist-in-training, on his way of being and his way of doing whilst supporting the protagonist's process.

The complexity of the supervisor's role is linked to each of the dimensions he warrantees: relational climate, triple focalization, multiplicity of processing functions, constructivity of feedback and methodology of processing.

Obviously each dimension is essential to humanistic processing and it is not always possible to develop each one. It is however a perquisite for the supervisor to closely monitor his role as facilitator, his first and foremost role, and to make sure that at all times he promote a relational climate characterised by the simultaneous presence of an unconditional positive regard, empathy and congruence.

These attitudes undoubtedly facilitate the security, confidence and freedom required to discover or expose oneself. They also seem to increase the tendency of the person to take responsibility for his/her life as much as he/she can do so. They clearly offer support and reassurance (Rogers, 1986, p.129): two essential elements of the process of change (Apter, 1999, p.78).

Whatever the scope of humanistic processing, the facilitator of a supervision session through psychodrama can thus not only encourage the trainee to gain some distance and to check his professional practice, but also to enable the learning community to develop its collective intelligence: through its manifold dimensions, humanistic processing then becomes a means to open up to new prospects, to put creativity to work, and thus to change through his ever evolving competencies.

Supervision is (mental) health promoting and stimulating to any professional, it can therefore but be of benefit to his clients. Everybody stands to gain from it: the professional himself who participates in supervision is neither dangerous nor endangered, he participates in a constructive dynamical development of his skills and competencies, helping him in turn to facilitate the development of others. No doubt, that was the real meaning behind my teacher's provoking statement.

References

Aguiar, M., & Tassinari, M. (1994). Processamento em psycodrama. *Revista da brasileiro de Psicodrama, 2*, 13-24.

Apter, N. (1987, 9 Février). Carl Rogers: Le deuxième souffle de la psychologie. *La Suisse*, 6-7.

Apter, N. (1996). A la rencontre de nos "ailleurs": Nos unicités. *Brennpunkt: Journal de la Société Suisse pour l'approche et la psychothérapie centrée sur le client* (69), 7-19.

Apter, N. (1999). Change: What is at stake? *International Journal of Psychotherapy (European Association for Psychotherapy), 4* (1), 67-78.

Apter, N. (2003). The human being: J.L. Moreno's vision in psychodrama. *International Journal of Psychotherapy (European Association for Psychotherapy), 8* (1), 31-36.

Bohart, A. C., & Greenberg, L. S. (1997). *Empathy reconsidered. New directions in psychotherapy.* Washington: American Psychological Society.

Buber, M. (1969). *Je, tu.* Paris: Edition Aubier.

Campedelli, A. M., & Wolff, J. R. (1980). Processamento de um onirodrama pela teoria da matriz de identidade de moreno. *Revista da FEBRAP, 3*, 78-79.

Fine, L. J. (1979). Psychodrama. In R. J. Corsini (Ed.), *Current psychotherapies*. Itasca, IL (USA): F. E. Peacock Editions.

Gardner, H. (1996). *Les intelligences multiples*. Paris: Retz Eds.

Goldman, E., & Morrison, D. (1984). *Psychodrama: Experience and process*. Dubuque: IA: Kendall Hunt.

Goleman, D. (1995). *Emotional intelligence*. New York: Bantam Books.

Karp, M. (1998). The director. Cognition in action. In M. Karp, P. Holmes, & K. Bradshaw Tauvon (Eds.), *The handbook of psychodrama* (pp. 147-165). London and New York: Routledge publ.

Kellermann, P. F. (1992). Processing in psychodrama. *Journal of Group Psychotherapy, Psychodrama & Sociometry, 45*, 63-73.

Leutz, G.-A. (1985). *Mettre sa vie en scène*. Paris: Edition Epi/Desclée de Brower.

Maslow, A. (1972). *Vers une psychologie de l'être*. Paris: Edition Fayard.

Mearns, D., & Thorne, B. (1999). *Person-centered counselling in action*. London: Sage Publication.

Monteiro, A. M., & Brito, v. (2000). Etica no psicodrama: Contextualizando o processamento. *Revista da brasileira de Psicodrama, 8*, 41-43.

Moreno, J. L. (1937). Interpersonal therapy and the psychopathology of interpersonal relations. In J. Moreno (Ed.), *Psychodrama* (Vol. 1). Beacon, N.Y: Beacon House.

Moreno, J. L. (1964). *Psychodrama* (Vol. 1). Beacon, N.Y.: Beacon House.

Naffah Neto, A., et al. (1981). Analise e processamento de uma sessao de psicoterapia psicodramatica grupal, cum protagonista. *Psicodrama, 3*, 3-16.

Rogers, C. R. (1951). *Client-centered therapy*. Boston: Houghton Mifflin.

Rogers, C. R. (1968). *Le développement de la personne*. Paris (France): Bordas/Dunod.

Rogers, C. R. (1972). *Liberté pour apprendre*. Paris: Dunod.

Rogers, C. R. (1975). Empathy: An unappreciated way of being. *The Counseling Psychologist, 5* (2), 2-10.

Rogers, C. R. (1986). A client-centered/person-centered approach to therapy. In I. Kutash, & A. Wolfe (Eds.), *Psychotherapist's casebook*. Jossey Bass.

Rogers, C. R., & Kinget, M. (1962). *Psychothérapies et relations humaines*. Paris: Editions Nauwelaerts.

Schmid, P. F. (2001). Acknowledgement: The art of responding. Dialogical and ethical perspectives on the challenge of unconditional relationships in therapy and beyond. In J. D. Bozarth, & P. Wilkins (Eds.), *Unconditional positive regard* (pp. 49-64). Herefordshire, UK: PCCS Books.

Williams, A. (1989). *The passionate technique*. London: Jessica Kingsley.

The Jungian approach: in situ supervision of psychodrama

Maurizio Gasseau & Leandra Perrotta

The authors present a Jungian psychodrama model of 'in situ' supervision where the supervisees have their work supervised on the spot in order to reflect on the multiplicity of processes inherent in a psychodrama group and learn from immediate experience. There are two co-psychotherapists in Jungian Psychodrama – a conductor and a narrator. The conductor directs the session whereas the narrator focuses on the group process, identifies the themes and restructures them into a newly integrated consciousness.

1. Different levels of supervision

Supervision is a fundamental experience for psychodrama trainees and qualified psychodramatists alike. There are various models of group supervision using psychodrama. Those used in the Jungian psychodrama Institute of the *Associazione Mediterranea di Psicodramma* and in the APRAGI-*Psicodramma* are:

- Psychodramatic supervision of individual cases;
- Verbal supervision of newly established and ongoing psychodrama groups;
- Psychodramatic supervision of psychodrama groups;
- 'In situ' supervision of psychodrama groups.

This article will investigate 'in situ' supervision, a form of supervision in which supervisees take turns at being conductor and narrator and have their work supervised on the spot by the supervisor. The supervisees become protagonists in a psychodrama conducted by the supervisor in a play regarding their work. This type of supervision facilitates a reflection on psychodramatic methodology as to what interventions are more adequate in a specific group and what kind of difficulties may emerge during a session.

2. The Jungian psychodrama model of in situ supervision

There are various candidates for in situ supervision:

- newly qualified psychodrama psychotherapists who have completed 1400

hours of training in four years with their trainers, plus 600 hours of supervised practice;
• psychodramatists who have completed their training but have not yet managed to establish their own group and would like to improve their skills in a safe space where they are encouraged to reflect on their mistakes;
• psychodramatists who need clinical supervision for their groups or who wish to grow professionally through permanent training by focusing on how to implement new and more efficient psychodramatic strategies in their work;
• third and fourth year psychodrama trainees who are training to conduct in advanced groups. Third year students will have completed 600 hours of clinical training, psychopathology, group psychology, group dynamics, analytical psychology, case supervisions, 2 years of psychodrama self-experience in an ongoing group, and will have undergone personal therapy. Third and fourth year students receive 90 hours of supervision in situ during their training, divided into 16 units of 5 hours each.

The ideal composition of in situ supervision groups is 8 to 10 supervisees, equally represented in gender, plus 1 or 2 supervisors. A supervision group lasts for 5 hours, divided into 2 units of 2 and a half hour each, with a 15 minute break in between units. Each unit includes a first session of 1 hour and 15 minutes conducted in turns by a different conductor, followed by a 10 minute narration by another supervisee. Following this, a 1 hour processing of the session takes place, using the grid below. The conductor and the narrator have their work supervised on the spot and are able to learn from an immediate experience and feedback. After the break, a similar session to the first takes place in which other 2 supervisees take turns as conductor and narrator of the group.

The decision to limit the first unit of the group to 1 hour and 15 minutes was due to the vast amount of material which needs to be analysed during the processing and because it is rather complex for inexperienced psychodramatists to conduct longer sessions. The frequency is determined by the objectives and the goals of supervision and by the institutional environment in which it takes place. The standard frequency however is twice monthly for psychodrama trainees or once monthly for psychodrama therapists using supervision as a permanent research group.

In situ supervision combines supervision of the activity, the acquisition of new techniques, reflection on the conductor's and the narrator's countertransference and the containment of the group. In the Jungian model of supervision, the supervisees who are not conducting the group take turns at being the protagonist.

In situ supervision is like a mystery to be solved by the supervisor: what profound motive moves the conductor to choose to play certain scenes? What 'shadow' prevents the conductor from carrying out certain actions? What unconscious fears, perhaps fruit of transgenerational echoes, influence the work of the psychodrama conductor or narrator?

When the supervisor directs the conductor in situ, images or memories emerge from the unconscious of the protagonist which often clarify the work.

The current model of in situ supervision was developed in Italy during the seventies by the French Lacanian psychodramatists, Gennie and Paul Lemoine. In their model, influenced by the work of Anne Ancelin Schützenberger (1998), there was a conductor and an observer in every psychodrama session. Their supervision method was very similar to the one developed and used in Italy in the late seventies by the Jungian psychodramatist and analyst Giulio Gasca (2004).

Jungian psychodramatists prefer the term *conductor* to the term *director,* used in classical psychodrama, because it refers to a group-analytical concept in which the setting is less theatrical. There are no real theatres or stages, nor instrumentation, props, masks or lights in Jungian psychodrama, only a round circle of chairs and a soft carpet on which the psychodrama takes place.

3. The theory of Jungian psychodrama

Jungian psychodrama (Gasseau & Gasca, 1991) is an analytically oriented group therapy, articulated in reference to the important innovations and concepts of J.L. Moreno in psychodrama such as tele and co-unconsciousness, to C.G. Jung's approach to depth psychology on dreams, collective unconsciousness, archetypal medicine and individuation and to Foulkes' concept of net and matrix (Foulkes, 1975).

Jungian psychodrama is a work of imaginative psychology in which the conductor's task is to gather and focus on the images which emerge from the group: these may include memories, dreams and actively imagined scenes. Role play is important in Jungian psychodrama, as is the representation of the imaginative realm. Particular attention is drawn to images which constitute a group emergence, containing prospectively useful meanings for the whole group.

Jung's concept of dream work is that the structure of dreams is similar to the structure of drama.

In Jungian psychodrama, dreams of all categories are enacted: symbolic dreams, visionary dreams, nightmares, oracle dreams, recurring dreams and social dreams. Dream incubation is a technique investigated by Gasseau according to the ancient ideas of Asclepius (Gasseau & Bernardini, 2009).

Dramatic plays, myths and rituals are interconnected in Jungian psychodrama. Dramatic plays contain rules (or messages) that rearrange and restructure the confused experience of suffering and the chaotic experience of the unconsciousness. Similarly, the ritual framework in Jungian psychodrama protects against the unconfined inflation of the unconscious. The protagonists in Jungian psychodrama are part of a mythical reality with universal meaning. This helps them to move beyond the experience of alienation into a dimension common to the whole of humanity.

4. The conductor

Conductors are told that they may decide for a 'time out', similar to the one in a basketball game, in which the group dynamic freezes and the conductor may consult with the supervisor on how to continue the psychodrama. Insecure conductors often need to be reassured on the validity and the usefulness of their interventions - such as playing another scene with the protagonist in association to the one played before, when to close a scene or when to use a specific technique. Sometimes a 'time out' is requested because the conductor doesn't know how to move forward and feels blocked and in need of reassurance and guidance.

The 'time out' lasts for a couple of minutes at the most and the conductor may then resume work with the protagonist. The supervisor goes back to a silent role and observes the scenes. Whenever the supervisor feels a serious omission in the scene may cause unnecessary suffering to the protagonist, due to a momentary impasse of the conductor, the supervisor may get up to double the conductor. The supervisor's intervention must be punctual, prompting the conductor to indicate a different possibility, such as "Maybe you could insert a helper. You could ask the protagonist whom he would like to have nearby in such a dramatic moment" or "Maybe the protagonist could meet his dead grandmother since she is the only one who can forgive him" or "Maybe you could play the unborn child".

Such interventions by the supervisor may confuse and destabilise the project of the conductor, so they are to be used only rarely and only in case of undue suffering of the protagonist. If the play is blocked as a result of a critical emotional state of the group such as inhibited aggressiveness, the supervisor may then double the conductor and suggest "Maybe the aggressiveness is too controlled. Try playing a scene in which the protagonist can express his anger." The conductor is free to act upon the supervisor's suggestion, having more information with which to decide upon a strategy. The supervisor otherwise never interrupts the conductors but leaves them free to experiment their own choices and particular style. Their work is of course later discussed in the processing of the session.

Supervision grid of the conductor

We have built a grid in order to facilitate in situ supervision of the conductor:

- Listening to the group
- Warm-up
- Selecting the protagonist
- Interviewing the protagonist
- Listening to the protagonist
- Scenic construction of the action
- Role reversal
- Doubling
- Soliloquy, mirror, role exploration and helper
- Length and closure of the scenes
- Scene resolution
- Virtual scenes
- Dream play
- Silence management
- De-roling
- Introducing the narrator

Listening to the group

Jungian psychodrama sessions in ongoing groups usually start in silence. The conductor must learn to respect the initial silence, manage the silences which occur between the communications of the group members and tolerate the emotions that can suddenly develop in this initial phase of the session.

A frequent mistake in analytical psychodrama is to invite the first person who talks about their problems or their suffering to become the protagonist, without giving adequate listening space to the whole group and to their personal themes and conflicts. It is important to train the conductors to listen to the whole group and not only to the potential protagonists. This allows the participants freedom of expression and a wider analytical breathing space.

Warm-up

As mentioned above, ongoing groups of Jungian psychodrama usually start in silence, with no need for a traditional warm-up. Listening to each member talk about their recent life events and personal issues is a warm-up in itself.

A more traditional warming-up of the group however is necessary with newly established groups in order to create mutual trust and group cohesion or with specific populations such as adolescents or borderline/psychotic patients.

Conductors may also choose to warm up the group if it has more than 10 members, if many weeks have passed since the last session or if the group energy is blocked: it then becomes necessary to reactivate spontaneity and creativity by warming up the group matrix.

The supervisor takes notes of the verbal indications given during the warm-up and signals any confusion or paradoxical communication in the messages, any incongruence in the sequences, the conductor's tone of voice, the tempo of the indications and any impossible tasks (such as evoking memories without leaving enough time). The warm-up must not be too dissonant with the themes of the session. The supervisor then invites the other supervisees to give a feedback of the warm-up during the processing and discussion.

Selecting the protagonist

In Jungian psychodrama, there is frequently more than one protagonist in each session and they are almost never sociometrically selected by the group. The protagonist may intuitively be chosen by the conductor if they feel someone is in pain or if the conductor senses that a group member has a particularly strong urge to explore a particular theme. Otherwise, the spontaneous criteria of self-selection is followed. The choice of the protagonist is a co-creation of the group, the conductor and the protagonist but it is important that the group emotionally approve the choice of the protagonist so that the psychodrama will be followed with greater attention and empathic participation by the group.

In this initial delicate phase, the conductor may suggest one of the participants play a scene based on a hunch, a feeling or a strategy. Jungian psychodrama consists in identifying images of memories or dreams and then playing them. Conductors must not be scared of playing scenes which are too painful or too dramatic nor must they be blinded by their own inner conflicts or by the problems of a potential protagonist.

It may sometimes be necessary to invite a reluctant protagonist to get up and play. The supervisor will observe how the conductor approaches the protagonist with words and gestures which encourage the protagonist in committing to a personal, analytical and psychodramatic work of individuation.

When the psychodrama of the first protagonist is over, the next delicate phase for the conductor to manage is the de-roling of the auxiliary egos, preparing the group for sharing and selecting the next protagonist. In this phase, the conductor may make the following mistakes:

- slowing down the work with the first protagonist, not separating from the protagonist's story or not closing the last scene for fear of having to start over with a new protagonist;
- choosing to work with the most cooperative and enthusiastic protagonist or with the easiest theme;
- not conceding a space and time of introspection to the group after the sharing so that a memory or an emotionally charged conflict may emerge.

Interviewing the protagonist

The conductor takes the protagonist by the hand and walks inside the analytical circle of the group while acquiring elements of the protagonist's current situation, past, affections, or fears. When interviewing a protagonist who is about to play a parent or a grandparent, it is important to verify if this relative is living or not. When listening to a dream, it is useful for the conductor to listen to its plot from the beginning, to ask the dreamer for details, and to try to visualise the contents of the dream.

If the protagonist is blocked or silent, or if no associations arise, the conductor must invite the protagonist to walk in the group in order to encourage body movement and reactivate the psyche.

The main mistakes conductors make in this phase are:

- placing themselves in front of the protagonists, hindering their visual field and blocking the flux of associations and memories;
- being too scared to ask the protagonists certain questions or of invading their personal space;
- hastily deciding to play a scene without sufficiently interviewing the protagonist first.

Listening to the protagonist

In a workshop during the FEPTO Annual Meeting in Bulgaria in 2002, the trainers were unanimous in declaring the importance of staying in touch with the suffering of the protagonist and of not escaping the pain. Avoidance is often the result of not wanting to contact one's own unresolved issues: un unmarried female conductor may have trouble playing a scene where the protagonist buys her wedding dress; a recent death in the family may block a conductor from playing a scene where a relative has died; those whose wish of having a child was not fulfilled may have trouble in playing a scene related to maternity. One of the fundamental tasks of supervision is to help the conductor discover what kind of choices were made and if there were any omissions.

When listening to the protagonist, the conductor must never lose contact with the group. Psychodrama must respect tempo and guarantee adequate timing to the protagonist and to the group. Even when the protagonist is on stage, the conductor must never forget the group and always be aware of any strong emotions or needs which must later be taken care of.

Scenic construction of the action

Viktor Frankl (1997) used to say that a psychotherapist cannot give a patient the meaning of life. The meaning of life is a profound 'Gestalt' developed by the patient: the therapist can only help the protagonist to identify and recognise it.

Jungian psychodrama is a group psychotherapeutic model with a strong attention to images and to the image process of the group. The conductor helps the protagonists play the internal images of their dreams, traumatic memories and affects. Scenic construction is a particular art and ample space must be given to the protagonists so that the images of their unconscious have the possibility of emerging.

In the supervision process, the supervisor often asks the conductor "What is the first memory that comes to your mind?" or "What is the first image associated to this event?" and the memory or image evoked by the psyche of the protagonist is then played.

The protagonist must be helped and encouraged to become the shaman who guides the group into the otherworld and into the images of the unconscious. It is important to build a scene which contains all the necessary roles and elements for its resolution. If the protagonists feel oppressed and the conductor does not help them to find and play the role of their internal oppressor, the scene will never be solved. If the conductor is unable to choose the right auxiliary egos or to place them correctly so as to presentify the scene adequately, the supervisor must then help the conductor understand what might have been necessary in the scenic construction.

Supervision is also helpful for shy or passive psychodramatists to teach them how to contribute to the scene and how to help the protagonist with their verbal directions: "Julia, you are a femme fatale and you are now moving towards Robert" or "You just fell out of bed and you are crying… soliloquy".

Psychodramatists are like orchestra directors who facilitate the movements, actions and interactions between the protagonist and the auxiliary-egos by guiding them rhythmically with their hands and facial expressions. Jungian psychodrama works on developing inner images, dreams, active imaginations and memories which are transformed into Gestalt images through the psychodramatic scene and the personal research of the protagonist. It is essential to choose those auxiliary-egos which will concretise and substantiate the memory of the protagonist. A dress hanging in a wardrobe calls for an auxiliary-ego to portray the dress, others to be the wardrobe and the unchosen dresses. It is just as important to play symbolic roles such as a wound, a door or a stain.

Role reversal

Role reversal is a fundamental technique in psychodrama. Supervisors must pay attention to which role reversals are necessary and which are superfluous. One of the tasks in supervision is to train psychodramatists in developing vital scenes which are not slow or boring. Time is an essential factor and it is important to learn to role reverse when necessary and to ask the protagonists if they feel they should role reverse with any of the auxiliary egos.

It is crucial to train supervisees on correct role reversal. When playing a highly emotional or dramatic scene such as the encounter with a dead relative or friend, the conductors may feel petrified in their movement and emotional plasticity, in resonance with the protagonist. But a dead person is not only a corpse: it can also become a spirit with which dialogue is possible. Role reversal is helpful to find messages in virtual scenes. Whenever there is a conflict or a fight, the protagonist must role reverse with the sparring partner. Even if the protagonists are lazily enjoying the sunshine lying beneath a tree, role reversing with the sun and with the tree allows them to decentralise their feelings and open up new points of view. A young woman who had been depressed for months, felt a strong sense of peace after role reversing with the tree under which she was standing.

Doubling

In Jungian psychodrama, the protagonist is doubled mainly by the conductor. Especially with psychotic patients or those with a fragile ego, the conductor must give words to thoughts, feelings and emotions which help sustain and empower the protagonist's ego.

The posture of the double is critical: with paranoid patients it may be beneficial to stand to one side so as to shield the persecutory threats perceived externally. Some psychodramatists imitate the physical bearing of the protagonist in order to double the inner psyche.

Often, supervisees lean their hands heavily on the protagonist while doubling or disappear behind the protagonist in the effort to identify with them. This does not allow for a decentralised position in order to resolve the scene. It is always best to alternate moments of empathic identification with the protagonist with a more external observation of the scene and keep a 'good enough' distance. Conductors who are too rational may double the protagonist repeatedly and excessively.

As for the contents of the doubling, supervisees must be taught not to whisper to the protagonist as the group must also be able to hear the insightful albeit painful echoes prompted to the protagonist when giving voice to thoughts and feelings.

Soliloquy, mirror, role exploration and helper

The soliloquy is a significant technique which allows the scene to be frozen into non action and gives the protagonist the possibility of entering the interstices of their own conscience. Young supervisees sometimes forget to use the soliloquy and prematurely double the protagonist, inducing themes and reflections active only in the mind of the conductor. Soliloquies help the protagonist achieve insightful thoughts on problems or shed new light on suffering. Soliloquies are useful after traumatic events, after an unexpected question or embarrassing request.

Supervisees must be made aware of how helpful it can be to invite the protagonist to observe the scene from the outside by choosing an alter ego as a mirror. This technique changes the observation point of the scene and allows the protagonist a new awareness without being overwhelmed by fear, impotence or sense of loss. The protagonists have the possibility of giving themselves a message and of making an active movement towards their own state of conscience.

Role exploration is a technique which allows the auxiliary egos to be more empathic. The protagonist introduces the different roles to the group but role exploration must be used sparingly as it may slow down the psychodramatic work.

The helper is a role which protects the protagonist during a difficult experience or traumatic event and helps them to feel less alone when playing painful memories. It is an empowering aid, a new role in the theatre of the mind of the protagonist. A sexually abused person can summon up the presence of a parent who was not there at the moment to protect. Interior dialogue with the helper is a resource which emotionally restructures the experience of impotence and abandonment and helps to build a new inner role of strength and resiliency.

A woman protagonist once played a scene in which she heard gun shots coming from her house in Sicily. She realised someone was being murdered by the Mafia in the room next to her. She was curled up in a foetal position and the conductor asked her if she needed anybody there with her. She knew her husband and her children were in the other room where her father in law was being shot and she was desperate for their presence near her. The husband and children were evoked on the scene and she was able to hug them and protect them and internally restructure her experience of helplessness.

Supervisees must be guided into using their counter-transference in order to sense when the solitude of the protagonist may become intolerable or when their need of protection may be inexpressible. Intuition of the feelings of the protagonist is essential in order to sense their deepest needs.

Length and closure of the scenes

Scenes must never be too long nor should conductors become trapped by their guilt because they don't know how to resolve the conflict. Scenes which drag on endlessly can become boring or superficial. But scenes - excluding vignettes - can also be too short if the conductor does not explore the roles adequately or if there are insufficient role reversals or soliloquies.

When a scene has expressed all its energy, it is advisable for the conductor to conclude it and open another. Conductors may be uncertain as to when to close a scene or may not know what to do in a moment of impasse. When there has been a catharsis such as laughter at the height of an emotional scene, supervisees must be taught to close the scene because the energy of the scene has dissipated. When all role reversals have been done and the play has nothing left to express for the personal development of the protagonist, the supervisee must be taught how to open a new scene based on the associations of the protagonist so as to tap on the transformative capacity of psychodrama.

Scene resolution

Zerka Toeman Moreno (2006) taught us that dramatic life experiences must be lived at least twice: the first time in reality and the second time in psychodramatic semi-reality. The scenes however cannot simply re-enact the dynamic but must be transformed and resolved.

It is important not to avoid the drama but to play it homoeopathically (Gasseau, 1995) or have the protagonist observe their alter ego play the scene. It is useful to teach supervisees how to introduce the technique of giving messages which activate communication among roles in the theatre of the mind of the protagonist, in order to achieve a deeper understanding. Messages convey unexpressed and meaningful contents to significant others. Scenes in which the protagonist is frozen or petrified on the scene, as in reality, may be contagious for the conductor who will end up being incapable of directing a scene in which transformation is possible. Correale describes a condition called 'memory hypertrophy' as a fixity of memories and a rigidity of roles which causes repetitive and obsessive behaviours (Kaes et al.,1998). When the protagonists are blocked, the conductor must help them contact their own inner subjectivity, not only by doubling them or by using soliloquy but by encouraging them into action. During the processing, the supervisor suggests how the conductor could have transformed the unresolved or blocked scene and considers possible alternatives. The task of the conductor is to help the protagonist recover communication when lacking, to encounter a loved one, to go into a protective situation of the past or to visit an imaginary happy place. In order to activate these scenic transformations, the conductor must be creative and master a vast psychodramatic repertoire of different solutions.

Virtual scenes

Virtual scenes are considered those in which events or relations have never actually taken place in life or in the dreams of the protagonist. These include meeting with an ancestor which the protagonist has never actually met – a key scene in transgenerational psychodrama (Perrotta, 2011) – or the encounter with a friend or relative who has recently passed away. The supervisor must observe how the conductor directs the scene and if the expression and the narration of the story are facilitated by the presence of these spirits. Supervisees are encouraged to be brave enough to play these scenes. They must learn how to facilitate the interaction among the various roles of virtual scenes, as if they were active imaginations, and practice facilitating the protagonist in answering questions or giving advice in the role of the ancestor. Role reversal is fundamental in order to identify with the psychic programming of the person evoked and to be able to talk in virtual scenes.

Supervisees must be helped on the timing of role reversal and on how to deal with the messages and the strong emotion the encounter with a dead spirit invariably causes. Particular attention must be paid to the way the supervisee closes a scene with a ghost and how to help the protagonist say goodbye. Deroling auxiliary egos from the role of a dead person is crucial and must be performed ritualistically and intensely.

A virtual scene can also portray the anticipation of an important event yet to come: a job interview or a declaration of love. Critical exchanges such as the communication of parents' divorce to their child or the infraction of a marital taboo to a spouse can be practiced and role played in a safe space.

Dream play

Jungian Analytical Psychology considers dreams to be the doorway of access to the soul and to the image reality of the unconscious. However, Jungian psychodramatists must be careful not to become overly fascinated by dreams, allowing the oneiric world to drape its mantle upon reality. A conductor may be seduced into playing the highly symbolic and archetypical dreams of an obese protagonist and overlook her problematic body dysphoria.

Psychodrama is one of the only modalities with which dreamers can walk in the landscape of their own dreams and have a perceptive experience of the dream images. Supervisees must be guided into helping the protagonist accompany the dream with all their psychic functions (Barz, 2009) so that they may feel with their senses, float in their free associations and help the dream to reveal its mystery.

Dreams have a dramatic structure and Jung believed that "the whole dream-work is essentially subjective, and a dream is a theatre in which the dreamer is himself the scene, the player, the prompter, the producer, the author, the public, and the critic." (Jung, 1934, pp. 149-172). When a man dreams of meeting a beautiful woman, symbol of his Anima, and of accompanying her into a forest and taking her into his arms, the dreamer must be invited to feel the warm hand of the woman in his perceptively, like in the dream. He must be able to smell her and play the dream with all of his senses.

Supervisors must be careful to observe the following things in dream representation: why wasn't the beginning of the dream played? Why didn't the protagonist role reverse with one of the auxiliary egos? Why wasn't the end of the dream played? In a particularly long dream, how could it have been divided into different scenes?

Oneiric scenes must be well staged and try to reflect the emotional richness of the dream itself. In the dream of a protagonist, her psychodrama group was enjoying a festive picnic in the countryside, eating fish and drinking local wines. The sky was strangely brightened by the simultaneous presence of the Sun, the Moon, the 7 stars of the Ursa Major and the North Star. Judge Giovanni Falcone, who had been assassinated by the Mafia, appeared and approached the group. Everybody was silent as the judge spoke: "I am Judge Falcone. I was murdered without being able to complete my work. Justice is important and you, social workers, physicians, therapists and teachers must continue my work." The rich 'Gestalt' of this dream had to be respected and auxiliary egos were chosen for the Sun, the Moon, the 7 stars of the Ursa Major and the North Star as well as for Judge Falcone and the group. The protagonist role reversed with each of the stars and was encouraged to give messages from the different locations in the sky.

Dreams give life to the dramatic 'personae' of our inner theatrical world and each of these represent a different philosophy present in the mind of the dreamer, different emotions, obscure feelings and 'shadows' which are hosted in the unconscious. Conductors sometimes forget to stage important roles. A conductor once overlooked the role of a stolen bag and the protagonist was left without the possibility of finding the bag and exploring it. In another dream, an abusive father had been kidnapped but the conductor had forgotten to choose an auxiliary ego for the missing father, thus depriving the protagonist of the possibility of a dialogue with him. In another case, the protagonist had found the key to a treasure chest but the conductor omitted choosing an auxiliary ego to play the chest and therefore the protagonist never got the chance to look inside the chest. Important symbolic elements can be found in certain roles which can give new meaning to the dream.

Moreno taught us to continue dreams and that nightmares were only unfinished dreams. When the dreamer is awakened by the terrifying contents of a dream, it is an unfinished dream, or as Grete Leutz says, a tragedy with no catharsis. The conductor must always invite the protagonist to continue the nightmare, allowing the protagonist the freedom of spontaneously and creatively transforming the end of the dream.

Silence management

Supervisors must draw their attention to the personal anguish evoked by silence: death anxiety, a sense of void, emptiness or feeling lost. Supervisees must be guided into silently listening to silence and to reassure the group with their presence. Silence is precious when it reflects a moment of introspection in the group, especially at the beginning of the session when group members often need a moment to themselves to get away from the chaotic outer world from whence they just came.

De-roling

De-roling is immensely important. It helps the auxiliary ego to shed the role taken on at the service of the protagonist. When auxiliary egos are called upon to enact intense and dramatic roles such as that of a suicide, a confused psychotic or a corpse, they risk absorbing elements of that role into their psyche. Ritual de-roling is the only modality available to protect the auxiliary egos from being subconsciously colonised by a foreign graft and dragging the consequences of that role into their own personal lives after the session.

De-roling consists in a ritual to help exit semi-reality. The protagonists place their hands on each auxiliary ego and tell them imperatively to quit playing that role. They announce they are no longer that persona and they repeat the auxiliary ego's name: "You are no longer my dead grandfather, you are Peter".

Rituals may differ slightly: Anne Ancelin Schützenberger invites the protagonist to slide their hands over the body of the auxiliary ego as if washing the role away. Some conductors have the auxiliary egos jump up and down three times while repeating their names or run around the room. What is essential however is the invocation of the person's real name and the invocation to quit the role.

Protagonists sometimes hug the auxiliary ego at the end of an emotionally charged scene, without actually de-roling them. Supervisors must point this out as the lack of de-roling can sometimes be devastating. Some auxiliary egos need an especially intense de-roling, especially when they have acted dramatic roles which resonate with personal issues or if they deeply empathised with the protagonist. Roles delve into the psyche, and can be healing or iatrogenic. It is the conductor's responsibility to protect the auxiliary egos, as well as the protagonist, at all times.

The whole group also benefits from a good de-roling. Scientific evidence has demonstrated the existence of a mirror neuron system matching action perception and execution (Damasio, 1999). Mirror neurons are increasingly being recognised as playing a fundamental part in interpersonal psychic processes by mediating the understanding of others' behaviour: to perceive an action is equivalent to internally simulating it. The same neural circuits are involved.

As Gallese posits in his work, embodied simulation is the functional mechanism at the basis of intentional attunement, our capacity to pre-reflexively identify with others (Gallese, 2005). Stricto sensu, mirror neurons tear down the self/other barrier and facilitate an embodied encounter with alterity. Thus, even from a neurophysiological point of view, the direct experiential link between agent and observer argues in favour of an adequate de-roling process.

Sharing

Sharing is the phase which follows the plays and is extremely healing: it integrates various therapeutic factors such as catharsis, cohesion, self-knowledge, existential factors, infusion of hope, universality, and therapeutic alliance with the psychodramatists and with the group (Yalom, 1985).

The plays are a gift the protagonist gives to the group and the sharing is a moment in which the others can give their own gift back to the protagonist.

The sharing must be well conducted and not left to the devices of the group. The auxiliary egos and the other members of the group must not analyze or interpret or exceed in role feedback but share their life stories or personal issues and show their empathy and personal attunement with the protagonist.

Introducing the narrator

Jungian psychodrama ends with the voice of the narrator. The transition from conductor to narrator must be smooth. The conductor must take special care in introducing the narrator but also in forewarning the narrator when it is almost the time for the narration. The final part of the session – usually ten to fifteen minutes - are devoted to the narration and the conductor can start to introduce the narrator with suggestive words such as "Let us hear a few more sharings and then listen to what the narrator has to say regarding the mysteries of this group".

Self-centred conductors sometimes forget to introduce the narrator and remain silent after the sharing or they devalue the work of the narrator by grumbling a few words under their breath about listening to the narrator.

Another mistake that conductors should avoid is to add their own point of view after the narration. Narrators have the final word and it is their task to find the fil rouge which connects the various scenes. The narration is a personal view of the group dynamic and hence never absolute and it is more reassuring for the group to listen to only one point of view rather than to have their co-therapists vie for the last word of attention.

5. In situ supervision of the conductor

In situ supervision takes place immediately after the session, with no break in between so that the tension of the psychodrama is still strong. The supervisor asks the conductor which themes they felt were particularly active in the protagonists and in the group. The conductor shares difficulties, criticalities and moments of impasse and how they worked through them. The supervisor will have taken notes during the session, based on the supervision grid shown above. The supervisor will also share any counter-transferential feelings evoked by the psychodrama and will suggest anything that might have been played differently. This part of supervision exposes the supervisor personally as well as professionally since the supervisor suggests which moments of the session were uncertain, which fears were blocking the conductor or the group, which depressive moments of stagnation were smothering the spontaneity and where the conductor had difficulty getting out of murky waters. In this phase of supervision, supervisees can take notes and this is also the time to clarify any technical, methodological or epistemological uncertainties.

The supervisor encourages the supervisees to respect their stylistic differences. Supervisors do not expect an exact replica of their own style but encourage the supervisees to explore their own personal and idiosyncratic modality of conduction.

Protagonist's feedback to the conductor

The supervisor asks the protagonists to give feedback to the conductors on how they felt during the psychodrama: if they were forced into being protagonists, if the doubling and the role reversals were pertinent and if the scenes expressed their inner reality. The supervisor also asks the protagonists to say if they had any special needs which were not met.

In situ supervision of the psychodrama: the conductor becomes a protagonist

In every supervision session, after the processing during which the supervisor points out limits and strengths to the conductor and after the protagonists' feedback, the supervisor invites the conductor to stand up and to play a scene. Psychodrama with the conductor as a protagonist helps the unconscious of the supervisee to produce an image or an underlying memory which clarify their work. The supervisor may sense a block in the protagonist, the fear of developing a certain scene or may perceive the difficulty in exploring themes of separation, intense sexuality or conflict.

The supervisor may ask "What is your first association to separation?" or "What is the first thing that comes to your mind regarding sexuality?" or "What do you associate to your parents' conflict?" and then has the protagonist role reverse with the 'personae' of his memory. This memory, in association to the previous psychodrama is a profound 'gestalt' which often reveals the difficulties experienced as conductor. Sometimes it is sufficient to ask the conductor "what is the first thing that comes to your mind when you think of the session?". It is a dialogue with the unconscious of the supervisee and the protagonist's associations will reveal any problematic knots.

The kind of associations requested by the supervisor are memories of the protagonist's daily life and not of dreams, active imaginations or virtual scenes which can distract from the real problems underlying the conduction.

In one case, the conductor's mother had recently passed away and she found herself with two potential protagonists in the group whose parents had just died. They both wished to work on this theme in order to cope with their grief but the conductor avoided the subject and decided to play a much less dramatic scene. During the in situ supervision, she played her own ambivalence about visiting her mother's grave at the cemetery. The supervisor associated her "not being able to go to the cemetery" to "not being able to play a scene with dead parents". The supervisee realised that she was unable to play the scene because her own psychic experience of the mourning was still unresolved.

In another case, the conductor avoided interviewing the protagonists and never asked them the necessary questions to understand their life experience so as to understand what scenes to play. The conductor seemed scared of asking questions and so during the supervision, when the supervisor asked her what she associated to this problem she remembered that as a child, she was often scolded because she asked too many questions and her parents would tell her to shut up. These imperatives had caused a fear of asking questions and the need to be silent. This attitude was not useful for a psychodramatist but her play as protagonist helped her to develop a new awareness of her personal declinations and of how to react to certain themes in the group.

The narrator is also invited to associate a personal memory to the material exposed during the narration and to play the relative association. For instance, a narrator with a decade-long experience was once watching a protagonist playing her panic-stricken self at the age of seven, watching her parents argue. The narrator suddenly closed her note book and blanked out. When it was her turn to narrate, she muttered a few things about the scene and completely forgot to mention that there had been another protagonist afterwards. When invited to be the protagonist during the in situ supervision, she remembered her parents fighting furiously and that she was very scared and had decided to hide herself in a big wardrobe so that she couldn't "see anything". And that is exactly what had happened to her during the narration. She had locked herself into an interior darkness which had not allowed her to see anything else.

In situ supervision is essential in order to help the conductor and the narrator gain new awareness of their tendencies to avoid or escape or forget. It helps them to confront crucial knots and problematic issues in a more mature way, in tune with the needs of the protagonist.

6. Narration in Jungian psychodrama

In Jungian Psychodrama, the narration is a powerful function of holding and containment. The final narration gives meaning to the session and is a vital therapeutic factor.

From a ritualistic point of view, the narrator is witness and protector of the group's transformation. The conductor alchemically dissolves whereas the narrator coagulates the dispersed elements and restructures them into a newly integrated consciousness. The narrator is the group's self-reflective memory and the narrator is a narrative-based medicine, a story that heals, identifies discrepancies and weaves the threads of the plot between past and present. The narrator gives voice to that which is unsaid or inexpressible.

One of the narrator's most important tasks is to identify the collective theme to which the group is reacting, the coherent mythical framework giving unity to the plays. The narrator opens up new perspectives and possibilities and plants the seeds generating creativity, allowing the new and the unexplored to enter the group. The narration is the space of memory, the historical and mythopoiec continuity of the group (Perrotta, 2009).

Supervision grid of the narrator

We have built a grid in order to facilitate supervision of the narrator in Jungian psychodrama.

- Identification of the collective theme of the group
- Synthetic skills
- Narrative and story telling skills
- Feeling skills
- Visual contact with the group
- Analytical skills
- Connecting the plays of the session
- Connecting to the previous sessions
- Use of counter-transference
- Missing elements or denial
- Amplification process
- Conclusion

The grids which have been introduced must not be followed rigidly but are a useful tool to investigate aspects and themes during supervision.

In situ supervision is a complex experience. The in situ approach of supervision in Jungian psychodrama presented by the authors establishes a model in which the conductor and the narrator have their work supervised on the spot in order to reflect on the multiplicity of processes inherent in a psychodrama group and learn from immediate experience. It is indeed an opportunity to explore issues as they resonate and are enacted.

References

Ancelin Schützenberger, A. (1998). *The Ancestor Syndrome: Transgenerational psychotherapy and the hidden links in the family tree*. London, New York: Routledge.

Barz, H. (2009). Psicodramma e sogno. In M. Gasseau & R. Bernardini, *Il sogno. Dalla psicologia analitica allo psicodramma junghiano* (pp. 269-287). Milano: Franco Angeli.

Clarkson, P. (1998). *Supervision: Psychoanalytic and Jungian perspectives*. London: Whurr.

Damasio, A. R. (1999). *The feeling of what happens: Body and emotion in the making of consciousness*. New York: Harcourt.

Foulkes, S. H. (1975). *Group Analytic Psychotherapy: Method and principles*. London: Gordon & Breach.

Frankl, V. (1997). *Man's search for ultimate meaning*. New York: Perseus Book Publishing.

Gallese, V. (2005). Embodied simulation: From neurons to phenomenal experience. *Phenomenology and the Cognitive Sciences*, 4, 23–48.

Gasca, G. (2004). *Psicodramma analitico. Punto d'incontro di metodologie psicoterapeutiche*. Milano: Franco Angeli.

Gasseau, M., & Bernardini, R. (2009). *Il sogno. Dalla psicologia analitica allo psicodramma junghiano*. Milano: Franco Angeli.

Gasseau, M., & Gasca, G. (1991). *Lo psicodramma junghiano*. Torino: Bollati Boringhieri.

Gasseau, M., & Scategni, W. (2007). Jungian Psychodrama: from theoretical to creative roots. In C. Baim, J. Burmeister, & M. Maciel, *Psychodrama: Advances in Theory and Practice* (pp. 261-269). London, New York: Routledge.

Gasseau, M. (1995). Lo psicodramma come trattamento omeopatico. *Psicodramma Analitico*, 5, 57-62.

Jung, C. G. (1934). *The collected works of C. G. Jung*. London, New York: Routledge.

Kaes, R., Pinel, J. P., Kernberg, O., Correale, A., Diet, E., & Duez, B. (1998). Sofferenza e psicopatologia dei legami istituzionali. Roma: Borla.

Moreno, J. L. (1987). *The Essential Moreno: Writings on Psychodrama, Group Method and Spontaneity*. New York, N.Y.: Springer.

Perrotta, L. (2009). Alchimia di sogni e rappresentazioni: l'osservazione nello psicodramma junghiano. In M. Gasseau, R. Bernardini, *Il sogno. Dalla psicologia analitica allo psicodramma junghiano* (pp. 338-350). Milano: Franco Angeli.

Perrotta, L. (2011). Introduzione. In A. Ancelin Schützenberger, *Psicogenealogia* (pp. 5-9). Roma: Di Renzo Editore.

Toeman Moreno, Z. (2006). *The quintessential Zerka*. London, New York: Routledge.

Yalom, I. D. (1985). *The theory and practice of group psychotherapy*. London, New York: Basic Books.

Interlocking gear wheels – from training to practice in various professional fields

Agnes Dudler & Kersti Weiß

In this chapter, we will describe and reflect upon the supervision process for trainees at the SZENEN Institute for Psychodrama. This process is an integral part of the overall psychodrama training programme in which we aim to train qualified psychodrama practitioners. The supervision process consists of two phases: The first takes place in the context of the ongoing advanced training group; the second is held outside the group, in the so-called practical phase, in which trainees apply psychodrama in their own occupational fields under the supervision of trainers.

Agnes Dudler, the director of the Institute as well as of the advanced training programme, is responsible for the first phase. In the second phase, Kersti Weiß plays a critical role for many of the trainees from SZENEN as the supervising instructor. Building on Moreno's wide-ranging perspective, Kersti Weiß developed a model (2007) which we have implemented in the supervision process and communicated to the trainees. This provides them with a tool through which they can clarify for themselves in what role, with whom, and in what environment – in what social, historical, cultural and political context – they are practicing psychodrama in connection with their primary profession.

The fundamental driving forces in the advanced training programme are self-exploration, methodological training, theory and supervision. These elements will shape the different phases of the training programme to varying degrees and with changing significance – and will continue to shape the participants' practice of psychodrama after the course has been completed.

Our goal for training and supervision is expressed by the words of a Tibetan Lama, who said: "If our pupils do not become better than ourselves, our tradition will die out."

1. Supervision during the training process

The psychodrama training programme at the SZENEN Institute for Psychodrama includes various forms of supervision:

- Supervision within the training group by the respective trainers or advanced training directors and other group members (beginning at the end of the first year of training);
- Intervision in peer groups (beginning in the 3rd year);
- Group supervision with specialised teaching supervisors (beginning in the 4th year);
- Individual supervision with a supervising instructor who has practical experience in the trainee's field of practice (beginning in the 3rd year as needed; otherwise beginning in the 4th year).

Supervision starts within the training group as soon as the participants begin leading small units on their own, at the latest in the second year of training. The peer counselling groups operate in parallel to this, usually beginning in the third year. Most of these groups remain together for several years, even after the advanced training is completed. In the fourth year, practical application outside the training group is accompanied by group as well as individual supervision. Here, the supervisors are trainers other than the primary leaders of the advanced training programme, often coming from other institutions.

Goals of the training programme and the ingredients for their success: The task of training psychodramatists who will enjoy their work and do it well in the long run – who understand the tools of their craft and can creatively tap new fields of application – is a multifaceted process which involves many different people. At the SZENEN Institute, it generally lasts around five years. The same rule can be applied here as to a well-prepared meal: The better the ingredients, the better the result. If you burn something or leave out an ingredient in the first stages of preparation, the final product will suffer. The better prepared the trainees are when they enter the external supervision phase the more they will benefit from it. Therefore, we place great importance on the selection of the group participants and the leaders. The guarantor of fundamental self-awareness is an experienced pair of trainers who consistently maintain primary leadership of the course, both together and in alternation, throughout the two years of basic training and continuing (with less frequent contact) until the completion of course. They are joined by other specialists who act as trainers in basic theory and in particular subjects. The older the training group, the more often the specialist trainers are involved. This stable framework ensures security, continuity and the flexibility necessary for experiences with specialised subjects, own experiments and other leaders. Thus, from the very beginning of the course, the participants are exposed to different models with which they can identify, making it easier for them to formulate their own roles as psychodramatists.

A group which remains stable over a long period of time fosters an atmosphere of trust in which participants can reveal their own weaknesses and learn to deal with them constructively. Even unsuccessful projects can be presented in the context of supervision, since there is a lot that we can learn from our mistakes. To this end, it is necessary from the beginning that participants are encouraged to both give and receive constructive feedback.

In order to cultivate the attitude of respect toward future patients, clients and customers which is fundamental to psychodrama, the participants are expected to be the subjects of their advanced training. We do not infantilise them and as much as possible, we avoid creating dependencies (for example, through individual teaching therapy within the institute). At the same time, from the very beginning – through the involvement of the body, the consideration of current political and social events as well as significant historical events (e.g., in sociodrama), a multi-gender and multi-generational perspective, and last but not least, a religious and spiritual dimension – we encourage a holistic view of oneself, of others and of any problems that arise.

With these intentions and ideas, we have developed a training concept in which the individual elements not only build upon one another in a linear fashion but rather – like the interlocking gear wheels of a clock – set each other in motion and drive each other forward. Even during the self-exploration phase at the beginning of the course, when the participants are still quite dependent, they can already sense the simultaneously turning wheels of autonomy and a feeling of community which will steadily increase in importance during the later phases of the process. At the same time, a solid and thorough exploration of oneself and good training in the first phase provide a fundamental and ongoing impetus for all the other gear wheels involved in professional development toward independent work. The entire process is also kept in motion by strong connections between the members of the training team.

2. The advanced training process

Experiences from many years of trainee supervision at various training institutions, as well as evaluation of feedback from alumni and guest instructors, have all contributed to the conception of SZENEN's current advanced training programme.

Year 1: Self-exploration: here, the exploration of the participants' own issues and those of the group are at the forefront. Through processing and reflecting on the leaders' decisions, trainees are made conscious of the techniques and approaches being used. During processing, the leaders reveal and explain what was done when and for what reason. In this way, the trainees become familiarised with the tools and fundamental processes of psychodrama in a playful and almost unconscious manner, through their own experiences, experiments and reflections.

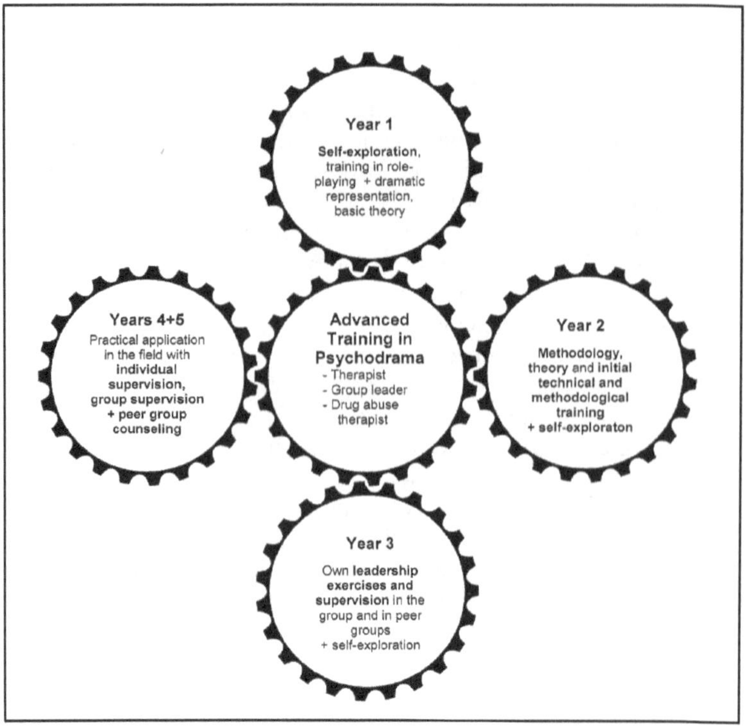

Figure 1: The advanced training process

The theoretical groundwork is taught in separate seminars and working group presentations (using psychodrama). Almost unnoticed, the trainees practice expressing themselves in psychodramatic terms: for example, when preparing short thematic scenes in small groups, or acting out or presenting important points from the warm-up: "Who are we as a group?" "What is the next step that a given individual/the group should take?" In this way, the trainees can practice diagnostic understanding and appropriate psychodramatic representation; they can carry out their first intervention planning in a simulated setting and observe the results in themselves and the other participants. The better each person comes to know him or herself, the more understanding and empathy she can develop for others; the less she will feel the need to hide and control – and she can thereby experience the freedom of revealing herself with all of her faults and weaknesses. These processes are fundamentally important for the participants' own practice and for the supervision process.

Year 2: Organic transition to more training and to praxeology, sociometry and group dynamics: technical training such as doubling, mirroring and role reversal is reinforced through the assignment of limited (focused) counselling or therapy exercises in pairs or small groups which are then reviewed and discussed with the group as a whole. Trainees can bring problems or questions from initial practical applications in their own professional fields onto the training stage and try out different variations under live supervision. In the evaluation process for the first two years – the basic training phase – small groups develop and direct a 'role atom'[1] for all of the participants. Thus, at this stage at the very latest, the participants are responsible for carrying out their first major task under live supervision – integrating practice in diagnosis as well as the description and development of personalities in psychodramatic roles.

Year 3: Individual practice and leadership in the group: exercises are repeated often in small groups, with the trainers serving as observers. The trainees are given therapeutic and counselling tasks to be carried out in individual and group settings. These are supervised live or subsequently by other members of the group, group leaders or guest trainers. This is the beginning of specific preparation to the individual role as director and the development of one's identity as a psychodramatist.

Parallel to the seminars, the working groups also complete the following tasks: in the 1st year, the trainees deal with psychodramatic theory and are introduced to it in the group through psychodrama; in the 2nd year, they cover group theories, which are also introduced in the group using psychodrama or sociodrama. In addition, trainees work more intensively on small practice exercises and unresolved topics from the training group. In the 3rd year, exercises and work on theory supplement the work in the group as a whole and peer group counselling on the trainees' own work in the field; trainees also direct one another in working through individual issues.

Years 4 and 5: Work in the trainees' own fields of practice: with individual, peer and group supervision as well as supplementary seminars on specific themes with specialised instructors. For his or her individual supervision, each

---

[1] The concept of the 'role atom', in German 'Rollenatom', comparable to the 'cultural atom', uses psychodramatic roles in relation to an 'inner director' to develop personalities (Dudler & Bosselmann, 2008).

participant chooses his own qualified supervisor – a person who has practical skills and psychodrama experience in the trainee's particular area of application. This portion of the supervision process is subject to confidentiality, even as far as the institute and the advanced training programme leaders are concerned. The institute will simply be informed that the supervision phase has been successfully completed or that an extension of the supervision phase is necessary. In this supervision context, trainees are able to develop and reflect upon their specific practical work in the field and have the opportunity to reflect once again on their personal set of circumstances.

Group supervision takes place in subgroups of approximately six participants in professional fields that are as closely related as possible or in fields of therapy, teaching or counselling that correspond to the degree being pursued. Pedagogically-oriented subgroups often choose to work with Kersti Weiß, who is known to them through two methodology seminars ('Crisis Intervention' and 'Sociometry in Group Processes') and who possesses a wealth of experience in this field. Therefore, in the following section, we have included her article on instructor supervision 'outside' the main training group or in a subgroup – supplemented by feedback from other supervising instructors and alumni of the psychodrama training course.

3. A sociodramatic concept and model of the factors affecting supervision

In order to explain my (K.W.) approach to supervision in the context of training for psychodramatists, I will first introduce the supervision model which I developed from supervision theory as well as from Moreno's teachings, to be aware of the cosmic and social influences at any given moment (Moreno, 1973, p. 3-8). For the training of psychodramatists, the model clearly illustrates the interaction of the many factors of reality and provides a basic orientation for the development of professional work as a psychodramatist or sociodramatist.

The outline of factors affecting supervision which I will describe below is intended to make accessible the complexity of the factors which influence a situation both from *within* and from *without*. Naturally, these factors are not only in operation in supervision situations: They are a part of everyday life and work. This model can help trainees to visually comprehend the way in which the many factors of influencing events are connected to one another, and where one might find new starting points for understanding and impulses for constructive action in a current or past situation. For supervisors, it can serve as a basic orientation point for developing a supervision concept and structuring the supervision process.

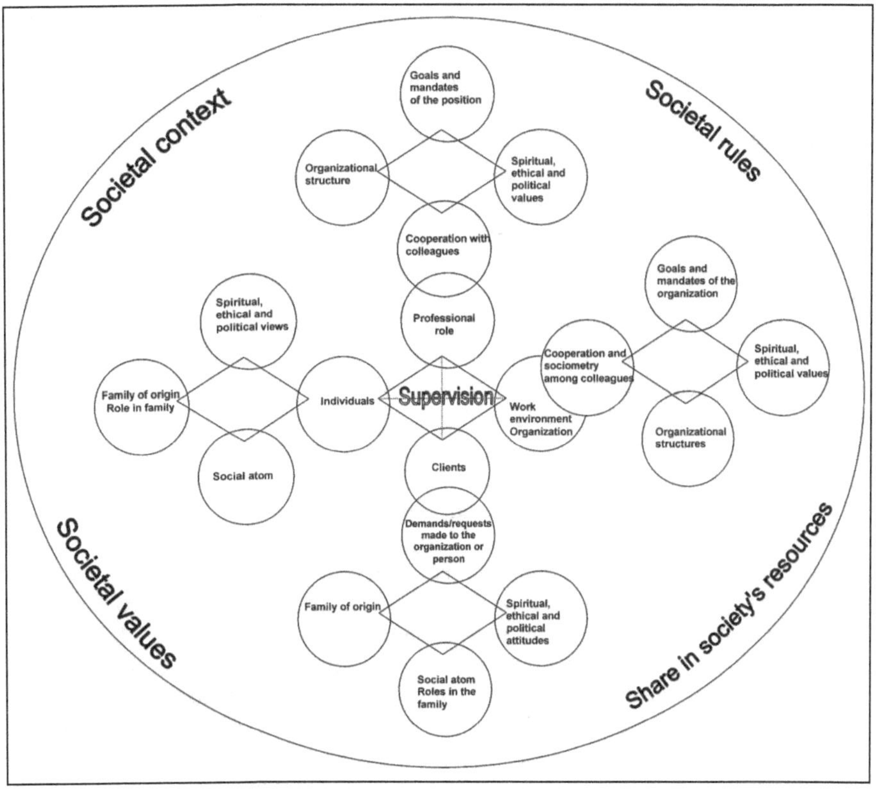

Figure 2: Factors affecting supervision (Weiß, 2007)

The model I have developed here is based on sociodrama, sociometry and psychodrama as a theory and methodology for action research – with the goal of applying them to supervision. The value of the approach presented below becomes particularly evident when we consider the fact that the demands faced by psychodramatists and supervisors today show parallels to the contemporary historical context and conditions in which Jacob Levy Moreno (1889–1974) formulated his conceptual interests and developed his theories (Buer, 1989, pp. 5-44). It was strongly influenced by a period of political and social upheaval, by Moreno's own spiritual and philosophical background and his widely varied professional roles and practical experiences. Moreno was interested in the comprehensive perception of the human being in all of his or her social, societal and cultural contexts. In his writings as well as in his practice, Moreno sought to understand people in all of their personal, psychological, social and universal behaviors and – with people as the subjects of his research – to generate possibilities for coping creatively with life (Buer, 1989, pp. 181-236).

By analysing the interaction between inner and outer reality and between psychological, physical and sociodramatic roles, we can systematically observe situational constellations. In the context of supervision and training supervision, this provides us with the opportunity to draw on J. L. Moreno's recognizable beliefs and on his seminal positions regarding people who come together in organizations.

The points of intersection between a person, his or her professional role, her occupational field, the organization and the client/customer on the one hand and the societal context on the other, as well as their respective interactions, are much more complex than simple individual experience (as inner life) initially reflects them to be. They have a lasting influence on individual perception and processing. The causes of conflicts – and therefore, the possibilities for resolving them – often lie outside of the individual person. Thus, the search for the causes of entanglements constitutes a significant portion of the supervision process. When individuals have recognised and accepted these causes, they can once again become productive actors. If we primarily limit our perception to the clients' or customers' internal or interpsychological realities and/or those of the trainees, we psychologise and individualise the process. In other words: This generally reduces the complexity of the problems at hand and thereby works against the critical and clarifying impulses of supervision, psychodrama and sociodrama – and their goal of freeing up the participants.

This model illustrates the complexity of the factors at work. It is extremely helpful and illuminating when the members of the group choose at least one example of their own with which they then cast the roles and set the entire model into psychodramatic action. All in all, the model can be seen as a kind of radar system in which the current landscape and the constellation of events at a given institution, in all of its complex processes, are revealed and presented. In further steps, participants can examine sections of the model more closely – more or less 'zooming in' on the individual parts to study them more in depth. The great challenge here is at the same time the most exciting task – and one which re-sparks the enthusiasm and commitment of all the participants: that of recognizing the way in which the individual factors work together to form an interactive context. (Naturally, however, it will not be possible to examine every dimension in depth in every supervision process.)

4. The starting points and the central goal of training supervision

People who choose to take a training course in psychodrama often have a secure professional identity and more or less extensive experience in their professional fields. The reasons for pursuing an additional professional qualification are manifold: the desire for a career change; having reached a crossroad of the professional practice, an expected, feared or hoped-for turning point in an individual's personal or professional life. Whether it is the wish to change, develop or discover something new or the motivation to acquire a vital new tool for one's professional practice – the desire for personal transformation almost always plays a central (if at first not entirely conscious) role in the decision.

The training programme and its supervisors are thus confronted with this tension between a desire for self-exploration (and the need to develop one's own personality) and the methodological and theoretical demands of the course. In most cases, the trainees experience this confrontation with themselves and with the reality of their lives and professional situations up to now as intense, dramatically revealing and, if all goes well – and it almost always does – as clarifying and finally satisfying. Despite all the theory and exercises included in the training, the self-exploration process, as a confrontation with the trainee's own self, has such a profound effect on the participants that nearly all of them begin the supervision phase with the concern that "Psychodrama won't work in my professional practice – never!"

This, therefore, is an essential goal: to successfully translate and modify the trainee's own important experiences from psychodramatic training to apply them in his or her own active practice (see also section: "Integrating one's professional and psychodramatic identities").

Although in both its concept and its execution, the training programme described above was largely developed with the goal of preparing trainees for this very process – meaning that the major protagonist work in the self-exploration phase should not be emphasised too strongly – most of the trainees, in fact, more or less consciously believe it is exactly this self-exploration work which constitutes 'real' psychodrama, and they measure their own attempts at application against it. This is usually the first important theme that we deal with in training supervision – no matter what institution the trainees come from. Other teaching supervisors confirm that in the supervision phase, it is usually necessary for trainees to relearn and adapt their skills to shorter interventions (compared to those experienced in the training group). Alfons Rothfeld gave an important contribution for an evaluation when he stated:

The supervisees were quite adept at applying the basic techniques of protagonist-centered psychodrama (scene setting, doubling, mirroring, role reversal, auxiliary ego appropriately. They arguably placed too high a value on the model of extensive theatrical/staged work in a long time frame. Thus, the necessity of adjusting to shorter periods of time with more vignette-like processing structures seemed to them to be a compromise or a rudimentary form of psychodrama (Rothfeld, August 2010, personal report).

We can understand this situation from at least two points of view: The profound experiences in the group processes, which led to personal growth and change, have left a lasting impression and the desire to work in the same way. Nevertheless, the conditions in the trainees' own work environments are usually different from those in the training group. Furthermore, given the skills that the trainees have developed up to this point, very few of them could imagine carrying out this kind of work in their accustomed professional roles. Thus, it is understandable that many trainees are reluctant to begin with the practical application of psychodrama if they have not already used it in the course of the last two years.

It is therefore important to work through and clearly define the following points with the trainees:

- What is my task, what is my role, and who are my clients?
- What does it really mean to work with psychodrama in my professional field – taking into account the setting, the institutional framework, the clients/patients/customers and the attitudes of my colleagues and superiors toward psychodrama?
- What is different here than in the advanced training group?
  At this point, it is encouraging to consciously remind oneself again of how effective even small exercises could be in one's own self-exploration phase and to attach value to those as well.

The journey through training supervision - parameters and mandate

The occasion of meeting one another in the supervision group provides an opportunity to experience and talk about the altered setting and the new phase of the training process. Here, the following elements come together: the trainees with their wishes and fears, the demands of the training curriculum and the supervisor with his or her possibilities, viewpoints, values, skills and knowledge. It is important here that once the parties have met and established a trusting relationship, they begin to negotiate exactly how the conditions of the supervision will be delineated, what content, requirements and themes should be focused on and given space and what should not. It is worth taking the time to discuss these issues, since their clarification will establish a secure basis for the quite unsettling phase to come: 'probation'. One alumna commented, "The informal working atmosphere in the group supervision phase was helpful as far as making it easier to discuss the supervision cases."

The training supervision process for future psychodramatists supports dimensions of discovery and learning at various levels, particularly with regard to

- the correct application of psychodramatic skills,
- their appropriate application to the desired task,
- integration of the practitioner's established professional identity and his or her new identity as a psychodramatist,
- theory and philosophy,
- methodology and research approach.

Although all of these dimensions can be found even in small interactions, during the supervision process we examine them specifically and individually in the context of the model presented above.

"Will psychodrama work in my professional environment?"

The candidates who continued the programme as far as the supervision phase beginning in the 4th year of training were poised to begin using the psychodramatic skills that they clearly possessed either within or outside their professional fields. I find it understandable that issues of self-exploration and professional identity crop up again at this stage; but it was possible to follow up on them in small increments as material for the reflective phase of the supervision process. Their training up to this point had definitely motivated the trainees to carry out their own psychodramatic work. Even though supervising instructor Alfons Rothfeld (personal communication August 2010) considers the supervisees to be ready to use their skills in practice, there will still be hurdles – both internal and external – which they will need to overcome along the way.

For example, the trainees often find themselves in the position of being the first person in their respective institutions to work with psychodrama – and thus having to introduce it, even though they themselves are not experienced psychodramatists yet. In the group, they have the opportunity to experiment with strategies for introducing psychodrama. Here, they can get a clear overview of their tasks, clientele, institutional framework and supportive as well as skeptical (or even hostile) colleagues and work through potential conflicts ahead of time.

The role-playing exercise is led by a member of the group and accompanied by live supervision. Here is one example:

A teacher has the job of teaching maths in a vocational training school class made up of students with learning difficulties. At first glance, both the teacher's task of having to submit evaluations and the previous negative experiences of her dispirited students seem to be at odds with her role as a psychodramatist. In the work 'on stage', the varied circumstances in her students' educational histories and those of the training group members are made clearly visible to the trainee. The process of working through their disappointments provides her with new insights; her newly-facilitated identification with the students sparks her creativity and spontaneity. She develops more and more ideas for psychodramatic maths instruction techniques through which she can communicate with and encourage her students. In the weeks that follow, the results of their learning are noticeably improved; the students are more alert and involved in the classroom – and their self-confidence increases.

Developing psychodrama skills

The trainees can also bring questions about their failed attempts at working with psychodrama into the supervision group. Together, the participants can use role-playing to examine and determine whether – and if so, when – something went wrong. This allows them to identify and examine processes of entanglement as well as to determine whether or not any 'small' technical mistakes were made or whether any serious problems will have a lasting effect.

For those trainees who work in individual settings, transposing the training material from the group into an individual situation often presents them with a hurdle. Here, creative trial and error in a workshop setting accompanied by a supervisor has proven to be a useful and necessary supplement to the earlier exercises.

Understanding the dynamics of a particular occupational field and developing a coherent diagnostic process and the subsequent ability to act upon it requires a great deal of practice and experience. We sometimes move forward in small steps, sometimes in great leaps – but we also sometimes go backwards. It is important to understand and accept this as a normal process of development and not become discouraged. While it is easier for some participants to speak about their mistakes in the secure atmosphere of individual supervision, experience from the supervision group – namely, that other people also make mistakes and have weaknesses and are not judged for them (as we also saw during the self-exploration group phase) – can be encouraging as well. Furthermore, discovering how much there is to learn from one's mistakes is also an important element of the learning process.

Role-playing exercises are generally led by other participants in the group. This leaves the supervising instructor free to assume other helpful roles – for example, as a double for the supervisee leader or as an expert during time-outs in difficult situations. We have also found it very instructive to incorporate group members as experts where necessary. Thus, during the supervision phase, the trainees use the skills that they have learned not only to help understand complex situations, but to continue experimenting and learning further. The act of reflecting on the 'case' and its circumstances as well as on the approaches used to examine it provides the trainees with an action-oriented understanding of the situation. At the same time, they are able to reflect on their leadership actions in the situation being used as an example, as well as in the current situation in the group.

Integrating one's professional and psychodramatic identities

The development of a personal identity as a psychodramatist is a central focus of the supervision phase. To quote Alfons Rothfeld once more:

Dealing with the impact of psychodrama training on the individual's own professional role, and his or her wishes for change and restructuring in the workplace reflecting his personal development through the training, accounted for a good portion of the supervision period (Rothfeld, August 2010, personal communication).

Having developed confidence in their skills and having had the experience of using psychodrama in diagnosis, intervention and reflection, trainees will be increasingly eager to apply what they have learned in their own professional fields. Thus, each trainee's future role as a psychodramatist needs to be integrated into his or her main occupation as a teacher, psychotherapist, personnel development specialist, social worker, minister/priest, physician or organizational consultant. This goes hand in hand with a relevant connection between the profession's basic orientation in its respective field (i.e., education, psychology, sociology, medicine, etc.) and the theory and practice of psychodrama, sociometry and sociodrama. This is where Moreno's overall theories become exciting.

As Moreno already discovered, philosophy, social theory, sociometry and role theory can, in their many fields of application – from theatre to organization and from psychotherapy to social policy – inspire individuals to expand their professional roles or redefine their professional practice as well as their scope for creativity. During the advanced training programme, most candidates develop a new and deeper connection to the theory and find it to be more grounding than they did before.

Thus, the training supervision phase is a setting in which the trainee can consciously observe, experience and reflect upon the supra-individual factors which influence the dynamics of his or her working environment as well as his clients and colleagues and his cooperation with them, and he can implement those factors in new ways in his practice.

Example 1: A teacher's profession is strongly defined by the obligation to impart specialised knowledge to her pupils and evaluate their progress – and a class is not a self-exploration group. However, if she can merge her professional self-concept as a teacher with her process skills as a psychodramatist, she can conceive of and develop her required tasks in an entirely new way: She considers both the individual pupils and the class as a group – along with the social and cultural values and networks that are present in the class, in the unique setting of the school system and in texts and other specialised tasks – as factors which influence one another; and along with the development of creativity and spontaneity as the basis of effective and sustained learning, they find their way into the new lesson ideas that she tries out. Even the simple step of examining the group dynamics of a class in sociometric terms and guiding it along more constructive paths will expand the vision and motivation of everyone in the classroom far beyond the limits that they have known up to this point.

Example 2: A minister, who, because of an increased workload, no longer found time outside of his primary duties to practice his long-cherished activities in psychodrama, developed and tried out a psychodramatic structure for delivering his sermons. As a result, both he and his parish were instilled with new energy and motivation thanks to the vivid and animated way in which he was able to communicate his messages.

These are just some of the ways in which a new concept of professional action can combine with the further development of a person's professional skills and identity. Our mutual joy of discovery in implementing psychodramatic practices in both familiar and new areas of application is, in itself, a fruitful ground for motivating and energizing our participants to further develop their psychodramatic skills in their everyday lives. Together, we have made many new discoveries and set many processes in motion which make our work as supervising instructors both exciting and rewarding.

5. Outlook

The feedback from our trainees, trainers and supervisors is an important tool which contributes to the constant improvement of our curriculum. Learning from our graduates' experiences in the training course requires regular evaluation of the processes and feedback from everyone involved. Both of these things generally occur once a year. The feedback we receive from our alumni – particularly on the subject of supervision – is especially informative, since in the practice phase of their training as well as in their subsequent daily practice, they will become even more keenly aware of what they have learned and what is still lacking. In this way, they give us suggestions for expanding, restructuring or shifting certain aspects of the curriculum and attest to the value of others. This feedback indicates that former trainees often recognise and appreciate the effectiveness of even seemingly small psychodramatic interventions in their own fields only after a longer and more satisfying period of practical experience. And only when their beginners' fears and insecurities have diminished, do many alumni realise how many of the skills they have internalised and now intuitively have at their disposal.

According to our assessment, the different phases and forms of supervision make an effective contribution to new psychodramatists' success at establishing their own creative and solid psychodramatic practices. This is helped by the fact that both the trainers and the supervising instructors are authentic and convincing psychodramatists themselves. In the trainees' development of their own identity, these people serve simultaneously as models and as respectful and constructively critical counterparts.

The idea of being 'only a cog in the machine' would seem to work in opposition to an individual's personal need for recognition. But just like in the workings of a clock, it is the quality of every individual gear wheel that makes the difference. It is our experience as well that a good ensemble can achieve more than individual stars who outshine all the others. After 19 years of work in advanced training, we can say (even if we don't always like to admit it): Some of our trainees are better than we are.

References

Buer, F. (1989). *Morenos therapeutische Philosophie*. Opladen: Leske & Budrich.

Buer, F. (1999). *Lehrbuch der Supervision*. Münster: Votum Verlag.

Dudler, A., & Bosselmann, R. (1996). Neun wesentliche Dimensionen einer effektiven Psychodrama-Weiterbildung. In R. Bosselmann, R., M. Gellert, & E. Lüffe-Leonhardt, E. (Eds.), *Variationen des Psychodramas* (pp. 365-372). Meezen: Verlag Christa Limmer.

Dudler, A., & Bosselmann, R. (2008). Das Rollenatom – Psychodramatische Rollen in Aktion. *Zeitschrift für Psychodrama und Soziometrie, 7*/2, 244-259.

Moreno, J. L. (1973). *Gruppenpsychotherapie und Psychodrama.* Stuttgart: Thieme.

Moreno, J. L. (1974). *Grundlagen der Soziometrie.* Opladen: Westdeutscher Verlag.

Weiß, K. (2007). Über- Blicke (zurück)-gewinnen. Wirkfaktoren in der Supervision. *Zeitschrift für Psychodrama und Soziometrie* 2, 185-200. Wiesbaden: VS Verlag.

Weiß, K. (2008a). Orientierung - Supervision und Coaching mit Gewerkschaftsmitarbeitenden. *Zeitschrift für Supervision, 1,* 25-30.

Weiß, K. (2008b). *Dynamik knapper Ressourcen* - Handbuch für Führungskräfte, SupervisorInnen und OrganisationsberaterInnen. Kassel: Uni Press.

Weiß, K. (2008c). Soziodrama, Soziometrie und Psychodrama als handlungsleitendes Supervisionskonzept. *DGSv aktuell: Konzepte für Supervision,* 7, 41-44.

Weiß, K. (2010). Knappe Ressourcen – neue Herausforderung für die Supervision. *Supervision, 1,* 18-26.

Weiß, K., & Bosselmann, R. (2005). Auf die Perspektive(n) kommt es an. Rollenwechsel und Rollentausch in der Organisationsentwicklung. *Zeitschrift für Psychodrama und Soziometrie, 1,* 103-114.

Supervision as self-observation within a structured training

Giovanni Boria

This article describes supervision for trainees in the four year school of specialisation in psychotherapy of the *Studio di Psicodramma* in Milan, Italy. This process is one of the fundamental steps of the training syllabus for those who are learning to manage the professional role of psychodrama-psychotherapist.

During their training students are engaged in two types of supervision: Firstly supervision is focusing on the experience of directing the training group. Secondly supervision is provided in the context of external institutions in the field of psychotherapy: trainees professionally work with clients assigned to them by the institution. In this chapter I refer only to the first type of supervision, which is offered within the *training group*.

The supervision is arranged in specific steps aimed at facilitating and stimulating an attitude of self-observation in the role of psychodrama director in order to gain insight into recurring emotional and cognitive functioning. The didactic structure of the schools enables cooperation among students with different levels of their training to reflect upon their experiences in the role of a director in psychodrama. Furthermore the chapter will describe how video-recording is used to facilitate this learning process. In order to exemplify crucial passages of this process excerpts of the supervisees as well as of the trainers are quoted to show how this didactic structure works for the students.

1. Structure of a training group

In this paper I use the term 'supervision' to indicate that formalized time when a healthcare worker critically reflects on how to manage their professional relationship with clients, in areas of psychotherapy and training, in collaboration with others (peers or others with a greater experience). I will illustrate how this process takes shape within a four-year specialization in psychotherapy, theoretically and methodologically based on the epistemology of Morenian psychodrama. I will therefore discuss the supervision of trainees.

The postgraduate specialization to which I refer requires a commitment of two thousand hours of theory and practice within four years. The training programme is structured with reference to the following areas: group training (600 hours), theoretical courses (240 hours), methodological seminars (240 hours), clinical seminars (320 hours), training activities (600 hours).

The *training group* consists of not more than 12 people and is directed following the psychodramatic methodology in a space especially structured for this purpose (psychodrama theatre). It takes place half a day per week. It is composed of students belonging to the first, second, third and fourth year of specialization. It changes each year with the exit of those who have completed the training process and the entrance of first year trainees.

Trainees practice in the role of director, with the support of two trainers present in the ongoing activity. The training groups meet and work in theatres located in different areas of Italy (Campania, Emilia-Romagna, Lombardia, Piemonte, Veneto).

Theoretical courses allow a structured learning of theory based on group-psychotherapeutic approaches, with particular reference to the methodology of psychodrama. These courses are held residentially in the months of December and April of each academic year. During each of these months, the trainees gather for four days in a hotel and attend the different courses according to their year of training.

All trainees meet monthly for *methodological* and *clinical seminars* (discussion groups), divided according to their year of training.

Supervised practice takes place half a day per week within an institution that provides interventions of a psychotherapeutic nature. The institution can be located anywhere in Italy: this activity is possible when there is a convention between school and institution that regulates the way of working.

1.1 Didactic function of a structured training group

The particular structure of the training group offers – throughout the whole training cycle – the opportunity to experience phenomena typical of an open group and to learn how to manage them. Indeed, the inclusion of new trainees at the beginning of each school year tests new ways of relating with acceptance and finding useful technical solutions, while separation from those who are preparing to leave the group at the end of the training requires work to develop the experience of mourning, both in those who are leaving and in those who stay. Moreover, the change in the group atmosphere, consequential to change in the composition of the group, trains the students to understand the variability of the relational atmosphere and to find ways of adaptation.

This structure also offers the advantage of facilitating its members at taking responsibility for directing the group. In fact, belonging to different years provides a gradation of skills that allows the distribution of more or less complex tasks: people from advanced levels can be assigned more demanding responsibilities, while beginners initially start with easier tasks.

All activities of the training are videotaped, allowing the director/trainees to watch themselves afterwards, comparing their feelings while engaged in the role with the perceptions offered them by an audiovisual tool.

1.2 Responsibilities of the trainers

The two psychodrama trainers in charge of the group are placed in a position external to it. They are responsible for:

- ensuring a smooth organizational functioning of the group (to set the physical space; taking care of starting and ending the session in time; distributing the responsibility among the students of directing the group according to a calendar);
- directing the group at particular times of the school calendar (such as the initial and concluding encounters of the school year);
- providing advice and support to those who are preparing to take on the responsibility of directing;
- establishing a constant reference point during the trainee's direction;
- taking notes of their critical reflections of the trainees' work and sending them to the trainee via mail within the next few days;
- a critical supervision of the directions which took place in the last month and on which trainees have already prepared their own individual review.

1.3 Procedure of direction in the group

The weekly half-day group consists of two units (or sessions) lasting two hours each, with a half-hour break in between. The role of director is distributed among several trainees and involves the use of different skills related to the course level to which each person belongs. The calendar indicating the structure of the work planned for each unit uses different letters of the alphabet to assign a specific task to each student.

The letter 'B' indicates that the student has been assigned the task of initiating the group. He or she has 30 minutes in which to implement the activities leading to the identification of a protagonist, who will then be directed by another trainee indicated with the letter 'E'. The letter 'B' is assigned to the students of the first year, a few months after their arrival in the group, while the letter 'E' is assigned to third year students, who thus begin to experiment the direction of a protagonist. Therefore, the letters 'B' and 'E' are always combined in the calendar.

The letter 'D' indicates a session entirely devoted to working with the whole group, without focusing on the protagonist. This type of direction is entrusted to students of the second year, to whom the more demanding responsibility of accompanying a protagonist in the exploration of their inner world is not yet demanded.

Finally the letter 'F' is assigned to more experienced students, those in the fourth year. Here the student must decide what type of work is more appropriate to the needs of the group: conducting a whole session centring on the group, or planning a work aimed at identifying a protagonist and the subsequent exploration of aspects of their inner world.

2. Direction, recording and self-reflection

This formalized step of supervision is identifiable with the time in which the students are confronted by their supervisors about their actions as practicing director (this unit of formalized supervision is shown in the calendar with the letter 'A'). In the school structure described above this time is placed in a monthly session in which students are confronted by the two trainers and by their peers. I will deal with this later, after considering other personal and private moments designed to make the trainee ready to make best use of this time.

Taking turns: Waiting for his/her turn to direct the group is a commitment for the director who feels the responsibility of stimulating his/her imagination and training his/her mind to formulate working hypotheses to search for operational models and to seek advice from the trainers. In the initial experiences, the directors may concentrate on themselves making themselves unavailable to listening to others. But this is a temporary phenomenon that can boost awareness at the time of supervision.

Implementation: The experience of directing is a great emotional and cognitive event for the trainee. Often there is an imbalance between emotions and mental clarity but the repetition of this kind of experience enables the student to find their balance, safety and individual working style.

Audio-visual recording: While on stage, the director creates interactive situations and the camera records the sequence of events. The camera is equipped with a wide-angle lens and is placed in a discreet and remote corner that allows a panoramic view of the environment. Environmental microphones capture and record the sounds clearly, regardless of the spatial location of the sound sources present in the theatre.

The technical aspects of recording are carried out by the trainers. The store data is a DVD, which belongs to the director/trainee. All students sign a release form (definition of privacy safeguard requirements) authorizing the registration and establishing criteria for using the recorded material.

Annotations of the trainers: The two trainers follow the trainees in their role of director from the audience, and take note of their thoughts, reflections and comments. They refrain from direct interventions to the person directing, unless considered necessary to solve a deadlock or to avoid the group or the protagonist experiencing unnecessary pain.

When the work is completed, the notes written by the two trainers take on a unitary form and culminate in an overall, educationally stimulating feedback. They arrive via mail to the student within the next few days.

Post-session feelings: The director/trainees usually have an emotionally charged experience. Often their expressions show a general sense of dissatisfaction on how they believe things went. Less often they are pleased with what happened. These responses are charged with emotions that require the right distance for a calm, critical view.

The trainers, immediately following the trainee's performance, do not comment on what happened, nor do they dwell on an emotional state that needs settling: the explanation of their views would be inadequate (even when solicited by the trainee) as they could block the inner processes that are operating deep in the mind of the person concerned. The trainee is left in a state of suspension: this will induce him/her to enter a quieter movement of self-observation that will culminate in the time of the formal supervision.

Private viewing of the video: The vision of oneself on screen acting as director of the group makes distance unavoidable and forces one to take a point of view that offers an unusual visual and auditory perception of behaviour. For some trainees this experience is quite difficult at the beginning: they need time to reach a kind of understanding, confidence, tuning, between the internal representation of the self and the image offered by the audiovisual tape.

An important, and not a secondary benefit provided by the taping is that it can be played in private, away from prying or critical eyes, in a situation of intimacy which allows to freely express reactions that are difficult to express in the presence of others, such as shame for macroscopic errors, regret at not having taken an important clue into account, and so on. It is also possible to interrupt the viewing at any moment, and to resume the viewing according to one's willingness to grasp and accept what has been digitally taped.

The possibility of revising a scenic representation or listening to a particular sentence several times, allows the student to capture moments of his direction with greater clarity, facilitating the overcoming of exceeded projections, prejudices and crystallizations present in the mind of the director/trainee.

Critical synthesis: The mental work of the trainee who has dealt with the revision of his own direction is produced in a paper to be sent (via mail) to peers and trainers in preparation of supervision (indicated on the calendar with the letter 'A'). It is a work of self-reflection and observation which compares the living experience of directing with the perceptual input provided by the DVD and the written comments sent by the trainers: at this point of individual effort, the trainee gives complete form to his/her thoughts, in a paper expressing the mental processes revolving around the direction. This writing expresses a free flow of thoughts and does not follow a pattern.

3. Formal supervision

Supervision takes place once a month for two hours and regards the trainee's directions of the group. Roughly six trainees are supervised in this space of time, those who have sent their written comments to trainers and peers.

This activity is led by the trainers and involves both the director/trainees and those present. It creates a dynamic verbal exchange deepening the feelings of the people involved (*double* function), soliciting peers' perceptions about the director in charge (*mirror* function), obtaining new points of view through the implementation of short scenes with the use of role reversal (*perceptual decentralization* function).

3.1 The supervisor's role

The two trainers have the explicit initial responsibility of ensuring the appropriate use of time (two hours are a short time!) allowing sufficient space for educational priorities. They base their assessment on the evidence already in their possession: the memory of the performances which they witnessed, their notes sent to the trainee, the written response returned by the student. Requests that spontaneously emerge during the meeting are also taken into account.

The priority criteria refer to those aspects of criticism, suggested by the different directions, which should not be overlooked: confused or mistaken methodological aspects; personal ways of directing that should be stigmatized; interference of expectations or biases that pollute harmony with the group or with the protagonist; feelings of the director that suggest further self-observation; and so on. However, enough time must be left for operations aimed at reinforcing or emphasizing the positive aspects in presence and action to be found in each director.

The trainers indicate the priorities which are analysed primarily through the verbal exchanges of the peers: they assign an initial time to the student object of attention, and then stimulate an expressive circularity involving more people. They are also careful to create the conditions that give a concrete object to the speech, requiring a clear definition of the situation they are dealing with. Often they turn a verbal description of a relationship into a short scene that allows the director to be in the shoes of this or that protagonist, experiencing perhaps unusual viewpoints.

The trainers are therefore those who provide sufficient space for reflection on the 'issues' which they deem more urgent, being clear about the essential parameter of the time available; and in this space they lead an interpersonal dynamic of both cognitive and emotional nature, able to stimulate new clarity with the contribution of all. Added to this is the opportunity that they formulate the final refinements that can be taken by the students as methodological and/or clinical references.

3.2 The supervisee's role

The supervisee, *focus* of attention and care of the group during the time of supervision, should be considered the most 'qualified' person of the matter dealt with at that time. He is the director of that particular invaluable experience which the group is preparing to consider, and because of which he has suffered or experienced pleasure, beginning with the direction and throughout different moments of specific mental work (watching the DVD, comments of the trainers, their own written reflections).

The supervisee has offered the group what has been maturing within them until that moment: in so doing, they act as a 'starter' of an emerging group dynamic, with the contribution of feelings and views of peers and trainers that complete and perhaps correct the initial vision. It is this initial expertise that can build new perspectives and clarities, producing a personal and professional advancement for all (supervisees, peers, trainers).

3.3 The peers' role

The particular composition of the group, which brings together students from different course levels, creates a hierarchy of skills among those present. This means that whoever is ahead in the training reinforces their awareness of what has been acquired, if the group is dealing with most basic contents level; and will also be induced to give valuable contributions coming from this knowledge. Those who are at the beginning of the training feel that the intellectual debate is beyond their understanding when the level of supervision is advanced.

For first year students, these moments, which represent a full immersion in the issues of psychodrama, are usually experienced as stages of maturing a future understanding. In the beginning the students' feelings of confusion, that may characterize their experience, are balanced by the insight that comes from knowing that these complex ideas are still not their responsibility. But what is perceived settles somewhere in their memory and will be rescued and re-evaluated in future times, with the overall progress of their training.

However, beyond the technical and methodological aspects of the direction, there is one area where interpersonal relationships are depicted as an equal exchange between people: this concerns the feelings of those who have been directed as protagonists or as members of the group. The subjective feedback that everyone can produce is a contribution that is not to be missed in the process of supervision. This contribution can be expressed as the *double* of oneself (explanation of one's own experience as member of the group, as protagonist, as auxiliary ego) or with a *mirror* offered to the director (their own image of the trainee playing the role of director).

4. Example of a supervision protocol

4.1 Supervision of Dan`s direction

Trainers: Dan please remind your peers of the activities ('B') you directed two weeks ago.
Dan: Zac had left us just before starting the session, so after I asked you to stretch, I invited you all to formulate a message in your mind that you could send to Zac. Then I asked you to blend into the group and to sniff each other as if you were animals. Then I invited you to take someone else by the hand, to start a dance and then to swap pairs. Then you sat in a circle and I asked you to go close to another group member and to tell them how you felt in that moment, choosing the partner with whom you would have liked to continue the previous experience. Finally I asked you to think of an image that came into your mind, and then to give it a title. Mark with the title ('unscrupulous') was chosen as the protagonist.
Trainers: Who would like to offer Dan a comment?

Maria (second year student): I appreciated your sensitivity very much because we were all in shock by the sudden news concerning Zac. You had the sensitivity to create a bridge between us and the companion who had left, creating closeness, I felt it as an expression of great sensitivity. Later on though it seemed to me that than you were a bit scared. In your directions for the next tasks I did not quite understand what I was supposed to do, even if the playful climate allowed us to be less dramatic, but we were all rather tense. When you had us smell our companions, you insisted we do it for too long and at one point I felt rather uncomfortable. It's ok for a moment, but if it goes on for too long it becomes intrusive. Overall, I appreciated your ability to recover, to take matters in hand, and not to leave us alone.
Bice (third year student): I felt the reference to Zac was an activity of reparation: it was a great insight on your part. In your shoes I would probably have asked myself if and how to face this problem. I also felt uncomfortable sniffing the others, perhaps because we needed to enter such an experience more gradually or perhaps because it touches me quite a bit: it is a very intimate contact and it would have been interesting to give this activity more time.
Trainer: Dan, would you like to explain the thoughts that were going through you?
Dan: I was pleased to be acknowledged and to have insights that were suited to the climate of the group. I had some moments of confusion, especially in expressing the tasks clearly. I was able to feel more peaceful and freer than other times. I tried to be more responsible in what I was doing, to be firmer and to want to be there. The situation was totally unexpected and I had to improvise.
Trainer: Do you have any questions?
Dan: I was told to give more structure to my tasks: in retrospective, I realize that there was a lack of readiness to structure an intuition that I had and that I ended up proposing in a generic way. Bice told me that I ended well, and I'd like to understand what she meant.
Bice: Since you worked on saying goodbye, I would have asked myself how to structure it. I felt that what you did, you did smoothly. But I felt the connection with the task of smelling each other wasn't clear.
Maria: Maybe you could have tried to connect the activities you proposed, to make more sense. There should be a unifying thread of thought behind the tasks because that would make it easier to enter the situation. I appreciated that you made room for something quite difficult such as saying goodbye, but then you found yourself trying to make the climate lighter by proposing unrelated things.
Conclusive remarks of the trainers: Congratulations on having reviewed yourself very well, you were very disciplined and balanced in your self-observation, without being too negative or too self-praising. You managed to focus on the

things that worked and on those that were still problematic. Congratulations on your intellectual and emotional courage. We appreciate your initial intuition. Of course, this contingency of saying goodbye could have been structured in a more accomplished way. The mental operation you asked everyone to do could have been materialized on the scene. A very simple thing could have been to take a pillow representing Zac, to put it in the centre and say: "Zac is leaving: I want each of you to give him a message". What everybody thought was only represented in the mind but it could also have been played on stage. After that you could have investigated the situation further: you could have asked everyone to think of a person (referring to the present or even the remote past) to whom they could address a thought inspired by the mood created by saying goodbye to Zac. In short, saying goodbye could have been an opportunity for further developments which could have led to the identification with the protagonist. However you managed to direct the group and after a few moments of confusion you took the situation in hand. One last thing about sniffing, which is a relevant but delicate experience. There was too much distance between the atmosphere of saying goodbye and the sniffing. Having the participants sniff each other out of the blue can be very daring. We can get to smell each other only after a series of steps justifying the action.

4.1 Remarks of the trainers on Dan's direction

The trainers send their remarks via e-mail to the trainee. They write the exact time indicated by the clock of the task given (which is briefly reported and written in italics). Here is the text produced and then sent to Dan:

(17:12 - Start of registration)
17:12 - You have a very difficult and unexpected moment. Do you feel like improvising?
17:13 – *"I will leave you to feel whatever happens to you, without giving you any tasks..."*
This is already a task.
17.14 – *"Try to make a gesture ... as if to give a message to Zac..."*
Very good, very good intuition, we need this precisely: but then you could go the whole way and play the idea of an explicit message to Zac, and then improvise a subsequent step that leads to the identification of the protagonist.
17.15 – *"... if you feel like giving a message of love... of reunion..."*
Ok. Very good. You are sensitive and careful, but be brave and propose something more structured!
17:17 – *"Walk around and smell each other like animals..."*
It is not in tune with the atmosphere above, you take a totally different path.
17.20 – *"Take the arm of a companion..."*

Confused and uncertain task.
17:21 – *"... like a 'Waltz'"*
The task continues to be confused. The pairs on stage are doing different things, they are taking things into their own hands.
17.23 - *"...and to finish, go and form a pair..."*
Ok, but for what? To talk or to dance? You must direct the group dynamic.
17.25 - *"Choose a companion and tell him or her how you felt in these past minutes".*
Ok, very good. This was clear and adequate.
17.33 - *"...close your eyes... see the images... choose one..."*
Ok, good progress. It would have been better to turn down the lights.
17.37 - *"...When you've found one give it a title..."*
Ok.
17.38 - *"Say your title out loud, one at a time... each of you will indicate the title which has spiked your curiosity the most".*
You are guiding the group well.
(Marco is chosen to be the protagonist).
Good Dan, you found yourself in a difficult situation; you showed us, thanks to your empathy and your intuition, that you had the 'right' ideas to face things 'head on'. You lost yourself when you felt insecure. You need to trust yourself more: have a good journey!

4.2 Dan's remarks

These are the remarks sent by Dan, stimulated by the vision of the DVD and the trainers' remarks:

I improvised the whole time, forgetting the plan I had prepared: the atmosphere created by Zac's episode before the beginning made the situation more difficult for me. Then I had a good idea but it was hard to give it a clear shape in my head and once I tossed it to the group I wasn't psychologically ready to turn it into a well structured task. I felt all of my personal limits: a feeling of insecurity and trust in what I wanted to propose as well as a lack of practical sense of how to transform an idea into a structured task.

Then I took the group to a completely different place, something I had also done in my first direction. This is probably a modality of mine, changing abruptly whenever the situation gets 'murky'. I must learn to be braver and to stay in the feelings, even when they are unpleasant (frustration or a sense of inadequacy). The confusing tasks I gave had the pairs directing themselves at a certain point because they did not feel directed by me.

I feel I must reinforce my responsibility in being there totally, in consolidating my attitude, which is still developing, of giving myself more concrete roles of stronger presence and of feeling more authoritative. I believe I must trust my resources more and feel more present in what I do (for example, in this case I should have organized my thoughts better and transmitted my knowledge in a clearer way) instead of losing my strength by following the labyrinth of my thoughts (in psychodramatic language that would mean

unblocking my actor-ego and containing my observer-ego).
I regret having the tendency of blocking my use of proxemics: of blocking myself in fixed positions instead of using the theatre space in a more mobile way. I would have felt freer to use the lights and props better (since I was blocked in my position, I did not feel the need to create the right atmosphere at the right moment when I asked them to go back inside themselves and find an image). As for the rest of the time, I was even starting to enjoy myself and felt a certain satisfaction; the ideas I had on the spot started to take on a shape and a direction. I managed to take things in hand and I never lost my mind or my nerves even when things got complicated.

5. Notes on the process of self-observation

Self-observation requires a stepping-out that allows trainees to look at themselves as an object. A reflective commitment made too early (as may happen, for example, if the time of supervision is placed immediately after the activity) could provoke a premature emotional discharge or expressions of verbal/cognitive nature, reassuring perhaps but not very effective in terms of evolution.

Distance in time must be satisfied to avoid the natural tendency to focus on details subjectively perceived as dominant and to take a broader perspective that allows us to capture an object probably different from fragmented images, constructed automatically. This object is reassembled in time, by using memories and impressions, watching the video recording and referring to the notes of the trainers.

The interpolation of a buffer time between the direction and supervision keeps the trainee in a state of suspension, trains him to tolerate the anxiety of waiting for the authoritative assessment of the trainers and to trust the flow of events without being conditioned by the approval of others.

The purpose of a critical reflection of one's own work, attained in supervision, is to mature new points of view and new beliefs that take the place of existing schemes, prejudices and projections: to replace the contents of the mind, subjectively perceived as 'real', with new 'truths.' The contents that are grafted in the mind through self-observation processes have the strength of evidence that comes from something that is recognized as belonging to oneself. They are not experienced as external meddling, as authoritative as the supervisor might be. In this operative framework, the role of the supervisor takes on many meanings of the Socratic method (maieutics) while suggesting specific points of view associated with his teaching function, he is particularly committed to ensuring that the supervisee can meet a variety of stimuli that lead him to look at himself and to discover truths.

6. Feelings of the supervisee

I thought it would be useful to quote the reflections of trainees belonging to different years of the experience I have illustrated. I asked them to report their thoughts freely. Below you can read what they produced.

6.1 Giulia, first year

When I arrived in my school of psychodrama, I was not aware of how much attention and planning was involved in the supervision of trainees. I was rather disoriented and concentrated on how to direct psychodrama. It took me a while to understand the complex (but not complicated) structure of supervision, but it then began to feel rather natural.

The first element that obliged me to be aware of what I was doing was when I was asked to identify a plan to direct the group (a ‘B’, an activity conducted by first year trainees, which helps a protagonist to emerge. The protagonist will then be directed by a third year trainee in an activity called ‘E’).

Thus, one week before my turn to direct the group, I thought up a plan to support the experience I intended to propose to the group. This forced/allowed me to carefully think of what I intended doing and to formulate a proposal which I then sent to my trainer by mail. I quickly received feedback from them. I modified a few things based on their suggestions, sent the plan back to them... and voilà... I felt more relaxed about my first conduction.

The second element relative to the methodology of supervision that struck me was the revelation that the trainees' performance would be filmed and that we would be able to watch our own experience as directors at home. So, for my first ‘B’, I was ready with my structured plan, validated by my trainers and with a blank DVD.

Once our performance is over, we are in an interrogative emotional state. Did things go well? What could I have improved? Doubts and judgments crowded my mind. When I have time (we're all usually quite busy), I ask my peers for their opinion, who are always very generous (they're always on my side!). If I feel brave, I try to approach my trainers indirectly for a sign of approval or encouragement.

Watching myself on DVD calmly and dispassionately but still emotionally involved was for me a great step ahead towards self-awareness of my actions. I was able to reflect on my behaviour, my movements, my intonation and the expressions that I would never have noticed. I have a reduced capacity of self-observation while emotionally immersed in my conduction. This step has revealed itself to be fundamental for me, allowing myself to split between observer-ego and observed-ego and to develop my learning. This is the cardinal point of the whole process, and the whole structure of supervision is based on this split like a play of mirrors.

At this point another fundamental piece of the puzzle is added to our learning process and self-knowledge: our trainers' supervision. The trainers send the trainee their notes, a few days after the conduction, containing suggestions, considerations, critiques, encouragement and praise. Thus, until now, there are two lights shining on our conduction still lit up by inexperience: the light coming from watching ourselves in action; the light coming from an expert watching us. This is the best way to focus on our strengths/weaknesses with greater clarity.

By comparing our reflections and by putting several points of view together, the trainee receives a comment on the conduction. This document is sent to the school director, to the trainers and to the trainee's peers. Also in this case, the activity of writing helps to redefine our impressions, our feelings, our mistakes during the group activity.

For me, writing down my comments while still in the heat of the moment, after having watched myself and having received feedback from my trainers is a critical moment when I judge my directing skills in a punctual and sometimes strict way. I understand that there is still a long way to go. I dwell upon my mistakes, my limits and I linger upon my strengths and resources benevolently. My thoughts often go to my group which supports me and to which I am most grateful.

The play of mirrors tied to supervision is still not over: in a school of classical psychodrama, how could the voice of the group be missing? Here is where a new element enriches the process with another point of view: the group's. This is the third element which sheds light on our conduction. This precious activity is carried out by asking the director/trainee for observations on the conduction and to synthetically delineate any reflections to the group. Then two or three peers talk about how they experienced the conduction. The group always participates in a very authentic way, never banal. There is so much tele in a group of psychodrama!

This other possibility of seeing oneself through another's eyes (first our own eyes, then the trainers' eyes, then our peers' eyes) is always very emotionally and cognitively enriching. I still think of my initial disorientation with benevolence and tenderness, but more and more aware that my journey has just begun and that there are still many discoveries to be made. And I feel I am not alone on this journey.

6.2 Catia, third year

I would like to illustrate the stimulating aspects of the supervision of my work as trainee in the *Scuola di Psicodramma* as well as those aspects which I found most difficult:

Stimulating aspects: Taping my directions on DVD allowed me a clear vision of my strengths and weaknesses. Watching myself direct psychodrama allowed me to identify my weaknesses without having to rely on my memories only. I was able to listen to the exact words I used or to those of the protagonist. In the psychodramatic method, where words and gestures of the director (as those of the protagonist) are very important, I was able to practice my skills in observation on one hand and the use of my words in a clear and suggestive way on the other.

The comments of my two trainers helped me to perceive their different styles of conduction. Although the feedback I receive from my trainers is one, the two trainers, who may have quite diverse opinions, evaluate my direction independently. This has helped me to contact personal styles in directing psychodrama (deriving from their own different training, age and experience).

I was struck by my feelings. Whilst writing my own notes after the feedback I received from my trainers and after watching the DVD, it was easier for me to reflect on my personal feelings and not only on the methodological/technical aspects of my direction.

In the group discussion, the feedback of my peers and of the protagonist was very useful. The presence of my peers during supervision allowed me to ask for their feedback on the activities proposed and on how they perceived me in the role of director. This helped me to recognize the effectiveness of certain interventions as opposed to others. The feedback given to me by the protagonist helped me to understand if I was able to accompany him during the psychodrama.

Group supervision helped me to learn from my peers' experience, allowing me to gain knowledge thanks to them and to be better prepared at directing psychodrama. Although my understanding of psychodrama was limited in my first year, I started to learn different elements relating to ways of directing psychodrama, which were helpful to me in the following years, allowing me to have a good background knowledge in the third and fourth year.

Difficult aspects: I had to overcome the difficulty of watching myself on video and listening to the sound of my own voice. Watching myself on video was quite unnatural at first and rather embarrassing. I noticed that my peers also had difficulty in tolerating the sound of their own voice and tried to avoid this unpleasant mirroring. However, this helped me to modify aspects of my habitual posture (improved understanding of my place on the stage) and of the use of my voice (modulating my tones) – elements which are quite important in psychodrama.

Directing my peers in psychodrama did not make my task easier. While conducting a group of patients (outside the school, during my supervised practice), I felt more natural and agile. My peers, better 'trained' in group work, often have trouble in letting themselves go, in being spontaneous, and this makes their participation more schematic or reticent. Furthermore, the themes which emerge while directing my peers often tend to avoid personal or private contents; whereas this does not happen while directing my group of patients in psychodrama during supervised practice.

Another element which I felt problematic was the sensation of a judging, yet silent presence of my trainers.

7. Conclusive remarks

The trainees enter this learning mechanism with great sense of trust. They often manifest the usefulness of watching themselves as an object of observation. This watching and perceiving is enriched by the perception of the trainers and the group members.

The trainees' experience goes through self-reflection and crosses many mirrors which multiply points of view and which prove the many possibilities of reading a situation as an important reference point: "the eyes of the video which never interprets, judges or changes but simply gives a direct feedback" (Laura's words).The stories told about the paths in training show meaningful inner movements towards self-awareness and the gaining of new competences. This is exactly the objective of the formation process we call supervision.

References

Boria, G. (2005). *Psicoterapia psicodrammatica.* Sviluppi del modello moreniano nel lavoro terapeutico con gruppi di adulti. Milano: Franco Angeli.

Muzzarelli, F. (2007). *Guidare l'apprendimento.* Milano: Franco Angeli.

Supervision at 'La Verveine'

Pierre Fontaine

'La Verveine' is a group for training in psychodrama and through psychodrama founded in Belgium in 1963. It merged with the Brussels CFIP (Centre for Training in Psycho-sociological Intervention) around 1973, and continued its training activity within this structure.

Psychodrama and training in psychodrama at the CFIP-La Verveine group has some specific features. I would like to emphasise the following aspects:

- The importance of the group. The protagonist emerges and declares himself in the warming up of the group discussion, as the bearer of the group's central theme. This approach derives from Anne Ancelin Schützenberger, our first instructor (1958-1966).
- The fairly intensive use of doubling by the co-director and group members. Doreen Elefthery introduced this habit. We worked intensively on psychodrama with Dean and Doreen Elefthery from 1967 to 1973.
- The co-direction of psychodrama and training in psychodrama. We have seen this done in different ways by Anne Schützenberger (co-director/ assistant), Paul and Gennie Lemoine of SEPT[1] (who alternated as director and co-director/ observer) and Dean and Doreen Elefthery (joint co-directors).

Supervision at La Verveine began in the early '60s, when a number of us, as co-therapists, were starting to use psychodrama in our work in institutions for young people and were encountering difficulties. We thus met in small, leaderless groups where we shared our knowledge and ignorance, acting through scenes again and trying to improve them. As we later discovered, this was a form of peer group supervision.

In this chapter different arrangements of supervision in psychodrama training are presented and discussed. This will be done by going through the phases of the training: the therapeutic and experiential phase, the training group, the internship as an assistant, and the supervised practice as an independent practitioner. We will then reflect on the elements and philosophy of this training.

---

[1] SEPT: Société d'Etudes du Psychodrame Pratique et Théorique (Association for the practical and theoretical study of psychodrama), Paris.

1. The experiential and therapeutic group

During this first stage, prospective psychodramatists begin their training by experiencing a therapeutic group, which is open to a wide range of people seeking psychological development or psychotherapy. The candidates experience psychodrama and the group: for example, they discover what it means to be doubled as a protagonist or to receive the group's 'echo'. They participate in psychodramas as protagonists, as antagonists, as doubles, and as group members – and hence are therapists to one another.

New participants who have to enter into new roles receive instructions, such as: "As an antagonist, you are playing his father, as seen by the protagonist," or "As a double, you must stand behind the person you are doubling, keep an eye on him, and check how he reacts to what you say." Such instructions are given to all members of the group, whether they are patients or candidate psychodramatists, but the latter probably pay more attention to them.

The therapeutic group enables the psychodramatist's to learn the profession in a practical way, as well as enhancing empathy and the capacity to give others psychological help; however, it is not supervision.

Entry into the experiential and therapeutic group is managed on the basis of participation in an introductory weekend on psychodrama and through interviews with the directors of the therapeutic group. Admission is governed primarily by the candidate's desire to work in the field of psychological and relationship problems (the indication for psychodrama) and by selection for membership of the group.

2. Supervision in the psychodrama training group

During the first sessions of the year, the trainers lead the group in order to ensure a certain degree of cohesion and at the same time present their style. They then work in blocks of two consecutive sessions each lasting about one and a half hours: a psychodrama session led by a trainee followed by a 'processing', in which participants reflect on the work and go back over what took place.

2.1 First session, centred on psychodrama led by the trainee

Warm-up and emergence of a director: The group is firstly taken in hand and set in motion by the trainer(s), enabling the members of the group to catch up on one another's news both personally, e.g. "Yes, I saw my mother-in-law..." and professionally, e.g. "I've been offered quarter-time work with two psychodrama sessions in the institution and...".

The trainer then asks who wants to lead the session. Reactions can be along the lines of: "I don't feel ready yet" or "Not this morning, but I could be a candidate for this afternoon." During the warm-up, someone may also say: "I would like to bring along a problem and be a protagonist".

The group members sort things out among themselves. Decisions are usually made quite quickly: it rarely occurs that there is a lack of candidates, or strong competition. The group takes account of the need for people to take turns in different roles and may give younger members a chance. The trainer validates the proposals.

Forming the team: The session director who has been chosen in this way decides whether he wishes to facilitate the group alone or with a co-director. The usual style is co-direction (see the section on co-direction later on). The trainers stress that the choice of co-director should be made on the basis of effectiveness and complementarity, not out of kindness.

Following this, the two directors have to agree on the assistance that the co-director will provide. They decide how they will work together. "When I'm a director I tend to stay too close to my protagonist. I'd like to leave this task to you as the double, so that I can stand back a little and look after the group as a whole. We could try this out. Would this be alright with you?" Or: "This is the first session of the weekend, and I would like to be able to spend some time on the warm-up; if you could double some of the members and see that the protagonist really emerges from the group. But it might also simply be a matter of "Right, shall we get started? We'll see how things work out."

Before they start, the trainer asks the directors to set time limits. He may also say: "I'll give you a maximum of one and a quarter hours, and a bonus if you come up with a vignette." We regard 75 minutes as a normal amount of time and we invite them to act out short scenes or vignettes.

The psychodrama session directed by the trainee psychodramatist: The trainee leads the entire session along with the co-director: the warm-up, the play, and the group feedback. The trainer will not intervene during the session unless there is real danger of doing something wrong or traumatising the protagonist. Fortunately this is very rare.

In these training groups, the director/supervisee can ask for a break in the session by raising his arm. During this break, he can talk with his team-mate(s) aloud or quietly, or seek assistance from the supervisor. He can also hold a brief intermediate group feedback to check on progress. The psychodrama scene which was 'frozen' or kept 'warm in the hay box' then comes to life again when the director/supervisee raises his arm once more and ends the break.

2.2 Second session focused on processing

By 'processing' we mean an analysis of how the session went. How did we work and proceed? Or how could we have done things differently? This session lasts about one and a half hours, with the time being allocated as follows: 10 to 15 minutes for the directors, 5 minutes for the protagonist, and the remaining time for the group and for opportunities to act certain sequences again.

Initial feedback and reflection of the direction team on the session: In an ordinary psychodrama, after the session, the directors withdraw to discuss the session and make notes on it. In the training group, after the coffee break, we start by gathering together the team who facilitated, placing them at one end of the circle and drawing an imaginary wall that isolates them from the rest of the group, which listens in silence.

Before the processing questions, there is always an opportunity for verbal and non-verbal expression by the facilitating team about "How do you feel personally and as a team after the performance?"

After this the trainer can, in a new group, facilitate the work of the team by asking questions such as: "What do you feel pleased about, in yourself or in your partner? At what point did you feel best? At what points were you least satisfied? What questions are you asking yourself? Are you looking for better ways of doing things? Did you have choices to make at certain points? What were the alternatives?"

Processing by the group. The protagonist: We normally start with the protagonist. He is also in training. What does he have to say about the support and 'guidance' given for the problem presented? What felt good? And what felt less good? What were the most important interventions or moments for him? We are engaging in processing, but we do not wish to carry on with the psychodrama. Sometimes the protagonist says that he still feels too immersed in what he has just gone through, and does not want to talk about it yet. We respect this wish. We turn to the candidate director and talk about processing.

The group: What was the process in the group? In developing the focal theme? In the choice of protagonist? The cohesion created around the protagonist? The sharing?

The group members individually, who also think that they could be directors another time. What did they learn? What would they want to take further or do differently? We thus embark on another task: that of not only analysing but replaying, showing what we have observed and seeking, as I used to say, 'further and differently'. This is what we call the 'laboratory of human relations', a test bench using role play. We have described this way of working under supervision in detail (Fontaine, 1975).

2.3 The laboratory of human relations

Chemists working in the pharmaceutical industry create hundreds of products until they find one with the sought after therapeutic properties and acceptable side effects. In the same way, we psychologists seek to extend our ways of working and test out alternative avenues, and we have to practise trying things out casually in order to get used to doing this.

We make up sketches that can be discarded. We are not working with a mother who might be traumatised, but with someone who is willing to put himself or herself into the place and mindset of the mother for a quarter of an hour and tell us how he or she felt in this role, and we swap roles so that we can experience how it feels too.

From a technical point of view, we work with concrete situations, reduced to two or three characters, for a limited period of two to five minutes maximum, consisting of three or four specific interventions (if there are more they tend to be forgotten). Then we listen to the experiences of those concerned and the members of the small group who volunteered, and we try out new versions. The supervisor maintains the balance between action and words, between the 'us' of the group and the question asked by one person, here for example by the supervised co-directors of the first session.

Processing by the trainers: Throughout the processing stage the trainers function in some ways as an echo, if only by repeating certain phrases, but at the end of the session, they sum up what has happened and what has been said. In assessing the directors' work, we always start with their strengths and then talk about aspects and lines of enquiry that need to be taken further.

Reflections: As we have described it, supervision is therefore carried out by several people who take on different roles, working in a fixed order: director/supervisee, co-director, protagonist, group, trainers. We follow this order by deliberate choice. It is also worth noting that this processing of supervision is more oriented towards 'technique' and is more group-based than the supervision of personal practice, where more attention is paid to the client and to understanding his/her personality.

3. The internship

We give the name 'internship' to work done by a 'trainee' as an assistant to a trained psychodramatist. When a student reaches a certain standard in directing psychodrama in the training group, the assessment process may indicate that he is ready to provide useful help to a trained, practising psychodramatist, and continue his own training in this manner.

3.1 The transition from training group to internship

Usually, when a trainee starts a psychodrama internship, he or she does not leave the training group, but remains there for some time. This ensures that the trainee retains the support of the training group, with the possibility of talking with colleagues at the same level about what the trainee is experiencing.

<table>
<tr><th colspan="2"></th><th>Training group</th><th>Internship with a practising psychodramatist</th></tr>
<tr><td colspan="2">Purpose of the psychodrama group</td><td>To learn the psychodramatist's profession</td><td>To treat the patient or client</td></tr>
<tr><td rowspan="3">TIME</td><td>Normal frequency of the group</td><td>Monthly</td><td>Weekly</td></tr>
<tr><td>Frequency of direction or co-direction</td><td>In a group of 8, one time out of four.</td><td>Always</td></tr>
<tr><td>Time between direction and echo</td><td>1/2 hour</td><td>Immediate</td></tr>
<tr><td rowspan="4">RELATIONSHIP WITH</td><td>Trainer</td><td>Parental transfer</td><td rowspan="2">Older brother, colleague</td></tr>
<tr><td>Co-director</td><td>Fraternal transfer</td></tr>
<tr><td>Client</td><td>• Colleagues
• Less severe problems
• Emphasis on the manner of conducting the processing</td><td>• Real patient
• Serious problems
• Emphasis on the case and its pathology</td></tr>
<tr><td>Third parties (paying party, doctor, family)</td><td>Absent</td><td>Present</td></tr>
</table>

Table 1: Transition from training group to internship

Of course, direction and co-direction of psychodrama continues, and is still done under live supervision, but even so there are many differences versus a psychodramatist's independent practice. We see some important elements that differentiate the training group from practice as a trainee:

• Firstly the aim of the psychodrama group is different: before, the group was set up to *learn* the psychodramatist's profession, whereas now it is working with the goal of helping, of treating, a client. This is an entry into real practice.
• In the training group, the drama's protagonist is a colleague who generally brings along *mild* psychological problems. The focus is on training in direction. During the internship in psychotherapy, the problems presented are *more serious*, and the client is there to receive help.
• The work in the internship is much more intense, usually a 3-4 hour session every week, for example, a group session and an individual session (monodrama). Slightly less time is spent on discussion, and there are no longer 8-10 people who wish to comment, ask questions or present or try out suggestions.
• In the training group of 8 people, each person only leads a quarter of the sessions as director or co-director, while during an internship the trainee co-facilitates or facilitates every session.
• In the training group, the trainer/supervisor is generally a *parental* figure for the transfer and the co-director is a *fraternal* figure. For a trainee in internship, the tutor often becomes a big brother, colleague, or team-mate figure.
• Finally, during an internship, it is good for the trainee to be initiated in and participate to some extent in relations with third parties such as the insurance system paying for the service, the referring doctor, or the patient's spouse. Thus the internship is undertaken within the real-life setting of intensive practice under live, close and immediate supervision.

To start the internship, the trainee must be able to take on the following roles:
• *auxiliary therapist, auxiliary ego, antagonist*, e.g. in individual psychodrama or as a professional double or as an observer giving his observations to the group at the end of the session or to the director in their discussion after the session;
• *co-therapist,* considered to be on the same level as the therapist. There are two ways of doing this: co-therapy with alternation between the complementary roles of director of psychodrama and auxiliary; or co-therapy with symmetrical sharing of tasks.

Being a trainee or assistant is like being an apprentice to a craftsman and seeing the master at work: watching how he succeeds and then hearing him explain his strategy. It is also important, say Napier and Whitaker (1972, p. 497), to see him fall from his pedestal at times: "He stuttered sometimes. His patients might leave him. He did not always understand. He wasn't always nice. What a relief that he was human after all…" The therapist is not just a mind that sees and thinks: he has his own emotional life, and engages with or withdraws from the group.

From the very first sessions the trainee is actively engaged. Some tutors leave it up to him to decide how much to be involved – to test the water before jumping in. Others put him to work immediately with a gesture: "Double here". Normally it is only after a trial period of getting to know one another, that the tutor lets the trainee direct part or all of the session.

The trainee receives feedback on his interventions fairly quickly. If he is facilitating together with the tutor, as in a dance, the trainee will feel his movements being followed or counterbalanced, adapted or deviated. In the discussion that follows, the trainee can talk about his experiences while they are still fresh, and receive feedback on the individual and joint work.

The trainee can also ask questions, including about the tutor's work. During the session he receives and perceives the group's reactions to his interventions during the warm-up to the drama and the feedback. There may be group dynamic effects, as in a long-term group, with relationship consequences on the group arising from the departure of the previous trainee and the arrival of his successor.

3.2 The relationship between trainee and tutor

The trainee must find or choose a tutor whom he can respect as a person, as a psychodramatist and as a trainer: as master, model and mentor, as Marineau (2006) puts it. There must be agreement about mutual expectations. These expectations must be expressed and heard. What help is expected from the trainee during sessions, how might this change with time, and how should this development be regulated? The trainee must say what he thinks, and the tutor will approve, give his support or baulk, and may reorient the trainee if necessary.

The trainee participates in a session discussion, receives feedback on his interventions, and can ask questions about the practitioner's interventions. Certain things need to be agreed in advance: How long can we or do we want to devote to this discussion after the session and possibly before it? What help is expected of the trainee on this subject? Should he make a summary of the session afterwards? Should he record the patient's progress? Should he take notes in his training journal or logbook about his own progress?

The trainee must accept the constraints of time and cost: transport arrangements and travel time, etc. Will there be some kind of remuneration? A balance needs to be struck between the trainee's input and the time that is devoted to him for training after the session.

The relationship between the practitioner or tutor and the trainee can be hierarchical and/or complementary: there is a boss and his assistant, or a trainer and a student. It can also be seen as an equal relationship. Even though one of the psychodramatists is more experienced than the other, they have equal rights – equal rights to be themselves and to be different during the session, and the freedom to work together.

This is a relationship in which the input of each is recognised, there is a division of labour, and there tends to be a consensus in favour of action. As a result, a team spirit develops that facilitates relationships: having a project to carry out together, being enthusiastic about it, trying things out together, moving forward, enjoying and suffering together, liking each other.

If things are not working out, we should perhaps not be afraid to get 'angry in front of our family'. Napier and Whitaker (1972) show this when one of them regularly arrives late for the session. Uphold the possibility of having different viewpoints, but maintain mutual respect. Therapists are thus in the position of models: they show that they too can have problems, but can talk about it. Pseudo-agreement would be more dangerous, concealing a struggle over differences.

The most important assessment to be carried out with the practitioner-tutor is that concerning the trainee's transition to his own independent practice under supervision. Although the trainee will see a supervisor before starting, getting the practitioner's opinion and discussing things with him are important, since the practitioner has an objective view of the trainee at work.

4. Supervision of personal practice

The trainee gradually reaches a certain degree of mastery of his working tool: his internship has taught him to meet and support patients, individually and in groups. If he receives a positive assessment, he can consider himself ready to start working in private practice, if the right circumstances arise, and to find clients or to be engaged as a group therapist in an institution. If he has found a team-mate, he now picks a supervisor.

This choice is important, even if the number of supervisors is locally restricted. Supervisor and supervisee need to know one another, to have already met, talked and agreed not only about the timing, length and charges for meetings, but also about the type of help that will be sought, and the supervisee's strengths and weaknesses. Will he be able to call on the supervisor in a crisis situation with a client? Is there a risk of a double relationship of friendship combined with authority and supervision?

4.1 Introduction to indirect supervision

Bernard and Goodyear (2009, p. 182)) noticed that those about to be supervised do not really understand what supervision is, and feel anxious about it. She had them watch a ten-minute video about what supervision is and how it takes place.

In our training, the point of transition from live supervision (within the training group and during an internship) to indirect supervision (which is re-counted, reported and recorded) of the supervisee's own practice is very important. To facilitate this transition, we can sometimes offer an introduction to indirect supervision. This is inspired by a simple experiment done by Mendelsohn and Ferber (1972, p. 443). It is practised in groups and takes place in three steps:

The *first part* takes place during the psychodrama session. This is led by trainees and takes place as is normal in the training group, except that one of the trainers is absent (and will be in charge of providing indirect supervision during the second part). After the role play, the directors talk over the session amongst themselves and prepare what they will present to the supervisor and ask him about.

Second part: supervision of the directors. This is indirect, since the supervisor does not know what took place. He is told what happened by the directors being supervised. In fact, he learns how they perceived the session and what they want to tell him about.

Mendelsohn and Ferber (1972, p. 443) were interested in "the inevitable distortion of therapists who do not do what they think they are doing", even while appreciating that "it is the perception of the experience, rather than what really happened, that seemed to be the real subject of the supervision."

The *third part* is the group discussion of the supervision process. The same question always comes up: can supervision be based on the perception of just one person? What importance should be given to the patient's experience (hence the usefulness of role reversal in supervision) and to the co-director's contribution?

4.2 Live supervision in the supervisee's own practice?

Is live supervision still possible later on, when young psychodramatists have their own practice? Some people feel that it is not possible because the presence of the supervisor would infantilise the therapist and clients would be reluctant to accept the interference of a third party.

We think it is possible to find suitable ways to continue with live supervision. We will give two examples. The first is that of a supervisor-'consultant', the second that of a supervisor-'assistant'.

Supervisor-consultant: I had a positive experience in psychodrama with people with mild mental disabilities. I was asked to supervise two young psychodramatists who wanted to start doing psychodrama at a day centre for people with much more severe disabilities. I accepted on condition that I could also attend sessions at their centre, if I felt it necessary. The group was therefore told that whatever was said and done would also be discussed with the Professor, and that he might come to meet the group, see what was going on and discuss it. I only went to the centre two or three times. The clients were happy that we attached so much importance to what they said and did, and the supervised psychodramatists and I myself found it useful.

Supervisor-assistant: One of our trainers worked in a psychiatric clinic and held weekly psychodrama groups with patients suffering from depression or psychosis. Caregivers from another psychiatric clinic were taking our psychodrama training and wanted to organise similar groups at their centre. They underwent a six-month clinical internship participating as assistants at sessions; they then requested live supervision at their own clinic when they began holding sessions there. The supervisor agreed to attend sessions conducted by the supervisees for a period of four months. The supervisor was able to intervene in the session by doubling and could thus draw attention to certain points. The supervisor intervened in the normal way, as practised at La Verveine, during the meeting following the session.

4.3 Pre-supervision

We believe that supervision should begin from the start of the group project. We then speak of 'pre-supervision'; Wilkins (1999, p. 41) refers to "preparing to work with clients or pre-group preparation". We tell our trainees that before the first psychodrama session in their own practice, a great deal of work must be done to lay the foundations which will in part determine what can be built up later on during therapy. This is because once choices have been made and announced, it is often difficult to reverse and change them. For this reason we believe it is desirable and necessary to start working under supervision on the very first steps involved in constructing the plans for the psychodrama group. We therefore talk over the practical organisation and the starting-up of a psychodrama practice. The subjects discussed are listed below. We ask therapists to read through the list (co-therapy) and see which subjects they would like to speak about with the supervisors.

This list offers many possibilities for choice, e.g. concerning the arrangement of the drama space for the group. We tend to let trainees develop their own style. We have already described this process elsewhere (Fontaine 2001, p. 309-316). However, we know that trainees tend to adopt an organisational approach similar to ours, even if they have observed several trainers.

Taking ownership of the space: We feel it is important for the directors and then the group itself to be able to take ownership of the space, even if it is used for other purposes during the rest of the week. Thus in an attempt to take ownership of a new room a director brought along a plain blue carpet, 3 metres by 4 metres. Before the session we would move the tables to the side and stack them, and unroll the carpet between the chairs. Together with the group we then had the sense: "Ah! Now we're at home!" And we could begin. At the end of the session, we would carefully roll up the carpet and place it, with a smile, on top of two cabinets standing side by side.

Rituals: As you can see, there are also opening and closing rituals, and not only those for taking ownership of a space. Fifty years ago we started doing psychodrama with adolescents with character problems. We would hold our session, at which they had the opportunity to express themselves and be themselves, and then they would be sent back to their respective care homes. The staff of the homes complained: "You give them far too much freedom; they are out of control when they get back". We then introduced a 10-15 minute transition period after the close of the session, asking them to help us give the room that had been made available to us a thorough tidying and sweeping, and we said that the authorities would be pleased and even amazed to find a room so well tidied. The adolescents then invented a ritual for the start of the session, with the aim of ensuring that what was said and done in the group would be kept secret from the staff of the homes. Together they mimed the famous three monkeys, saying: "We all agree: see nothing, hear nothing, mouths sewn shut or head cut off and floating on the water.[2]" This after a girl once told the director of the home, "Sister, we acted you in the psychodrama."

[2] We do not know where the last part of this formulation comes from. In Leuven, the Blessed Margaret is honoured, she was the proud maid at an inn who refused advances and was murdered (1255); she was then thrown into the water and her body floated upstream.

The place: • The psychodrama room: dimensions. • Will the stage be at the centre, or in the gap between the group seated in a horse-shoe arrangement? • Seats or cushions, carpets. • Props, small table, lightweight chairs, stool, cloth. • Sound or video recording. • Accessories for the break: glasses, drinks, tray + cupboard for storage. • Directors' room (for discussion and relaxation).	***The time:*** • Length of sessions + reflection time, discussion time for directors, before and after the session. • Length of course: a) in outpatient care: 1 session, 2 sessions, day, weekend, week, 2 weeks, commitment for 10 evenings, then a break; or b) in institute • Time between sessions • Duration of client's commitment/ lack of punctuality - wait until everyone is there? Or start?
The directors: • Alone or co-facilitating? • Agreement about time commitment outside the session. • Distribution of tasks and roles, complementarity/alternating in session. • Roles outside sessions: note taking/files / recordings/contacts with referring party. • Financial arrangements. • Authority and responsibility. • Team supervisor. • Who to consult if tensions arise in the direction team.	***Group of participants:*** • Characteristics of participants sought: age, gender; group: size, heterogeneity? • What is feared? • Announcement of the session/text • Selection, rules, procedure, • Initiation or awareness session(s), interview(s). • Procedure for entry, selection, group, test, interview, admission. • Procedure for leaving the group, warning, moments.
Social context in an institution: • Personal interviews and information about psychodrama, and directors. • Listening to requests • Informing and negotiating with management and staff about the group project (see above), place, time, group participants, financial issues. • Informing, listening to those concerned, e.g. nurses, individual psychiatrist, paramedical staff, assistants.	***Social context in private practice:*** • Information sheet for doctors, psychotherapists, etc. and their clients • What is psychodrama? • Presentation of directors • Practical details, place, time. • Financial and legal conditions. • Registration, admission.

Table 2: List for pre-supervision

5. Supervision after training and recognition

Although we know that training continues throughout our working life, there comes a time when we are recognised as being able to practise independently. Our training needs to gradually ensure a smooth transition between independence and a degree of interdependence. We will now briefly look at peer supervision groups and self-supervision. These are natural extensions of our supervision programme, in which the group is very important, but the primary focus is on the supervisee himself.

5.1. The peer supervision group

This group can be formed from a collaborative supervision group of the kind we have described, if five or six members of the group feel comfortable together and are sufficiently confident that they can learn to manage without a supervisor and share the responsibility amongst themselves. Others make up a group according to their geographical location.

The tasks within the group are passed around in turn. The following must be assigned every time:

- A case presenter, who presents one or two cases and asks questions for which the help of the group is sought.
- A moderator, who leads the meeting, determining who speaks and acts.
- The other group members are the consultants, who seek to find solutions along with the presenter.
- Sometimes we add an observer/ secretary, who at the end of the meeting can echo how the group has worked and what it has learned.

These groups do not just supervise cases: often a member wishes to present a new technique or theoretical standpoint. In these groups the members do not only get help for the work with their clients, but they can also strengthen social ties with their colleagues and benefit for their professional learning.

5.2. Self-supervision

Williams (1995, p. 212) says "the group works together to develop the role of 'self-consultant' in each member, so that each of them becomes a self-monitoring professional," and explains how the group acts (id, p. 211). Todd says the same thing about supervisors: "Supervisors are expected to prepare supervisees for self-supervision." (Todd, 1997/2002, p. 19).

6. Conclusion

Having reached the end-point, it would be honest, I feel, to ask myself the same question as our supervisees: What do you feel pleased about? What did you like about your way of doing things?

I would say this: I like self-supervision, for example the moment when the supervisee assesses himself, authentically and clear-sightedly, during the training process, supported by the respect and sincerity of his colleagues. It is a moment when he is truly himself, in connection with and thanks to the others.

I also like group work, especially the moments when, in a psychosocial workshop, the group seeks answers which are both appropriate and creative.

And lastly I like following trainees' progress. One supervisee said to me: "I came here for the first time six years ago, with many questions. I still have questions but I have learned how to think."

In these examples, whether synchronously or diachronously, I find traces of M. Buber, E. Erikson, J. L. Moreno and many other authors, with regard to the theme of what I consider to be a healthy individual, a healthy family, a healthy system (Fontaine, 1985). I would especially like to emphasise the amazing complementarity of 'being oneself' and 'being with others', as well as the permanence of continuity during the change process.

We should remember that Moreno did not just work with groups of patients. He was assisted by a direction team: "...how to get inside the psyche of the protagonist? You do not have to do that alone as a director, you use a member of your team. In psychodrama we work as a team" (Moreno et al., 2000, p. 69).

I have expressed my conviction and faith in co-training. I would like to remember and mention here those who were my main team-mates in the training groups: Nand Cuvelier, Dorothée Fontaine-Nyssens, Herman Engelhardt, Bernard Robinson and Chantal Nève-Hanquet. I also wish to thank Christine Vander Borght and Micheline Weinstock, with whom we collaborated in creating this text.

References

Bernard, J. M., & Goodyear, R. K. (2009, 4thed.). *Fundamentals of Clinical Supervision*. Upper Saddle River, N. J.: Pearson.

Fontaine, P. (1994). In Belgium CFIP – Verveine. In P. Fontaine (Ed.), *Study and Coordination of Psychodrama Education in Europe* (pp. 46-49). Oxford Meeting. 1994. Preliminary documents.

Fontaine, P. (1975). Structuration en jeu de rôle de formation. *Acta Psychiatrica Belgica, 75*, 905-915.

Fontaine P. (1985). Familles saines. Esquisse conceptuelle générale. *Thérapie Familiale 6* (3), 267-282.

Fontaine, P. (2001, 2nded.). From admissions to qualifications: the transitions in the psychodrama training curriculum. In P. Fontaine (Ed.), *Psychodrama training. A European view* (pp. 99-110). Leuven: (BE) FEPTO Publications.

Fontaine, P. (2001, 2nded.). The development of the trainee's own style. In P. Fontaine (Ed.), *Psychodrama training. A European view* (pp. 309-316). Leuven: (BE) FEPTO Publications.

Marineau, R. F. (2006). *Supervision in Psychodrama and group work and the model of the reflexive practitioner. A Summary*. Workshop at IAGP Congress at Sao Paolo.

Mendelsohn, M., & Ferber, A. (1972). Is everybody watching? In A. Ferber, M. Mendelsohn, & Napier A. (Eds.), *The Book of Family Therapy* (pp. 431-444). Science House. Aronson.

Moreno, Z., Blomkvist, L. D., & Rützel, Th. (2000). *Psychodrama, Surplus-reality and the Art of Healing*. London: Routledge.

Napier, A. & Whitaker, C. (1972). A Conversation about Co-Therapy. In A. Ferber, M. Mendelsohn, & A. Napier (Eds.). *The Book of Family Therapy* (pp. 480-506), Science House. Aronson.

Todd, T. C. (1997/2002). Self supervision as a universal supervisory goal. In T. C. Todd, & C. L. Storm, *The complete systemic supervisor. Context, Philosophy and Pragmatics* (pp. 17-25). New York: Authors Choice Press.

Wilkins, P. (1999). *Psychodrama*. London: Sage.

Williams, A. (1995). *Visual and Active Supervision. Roles, focus, technique*. New York: Norton.

II. Relational Aspects between Supervisor, Supervisee and Client

Relational aspects between supervisor, supervisee and client – an introduction

Pierre Fontaine & Jutta Fürst

As supervisors we relate not only to the supervisee but also to the client of the supervisee and furthermore to the emotional significant people the client is related to. It gets even more complicate if the supervisee is working with a group of clients and/or the supervisor is supervising a group of supervisees.

In supervision relationship is important, as it is in counselling and psychotherapy. We can say "In a pyramid fashion, the supervisory relationship is a relationship about a relationship about other relationships" (Fiscalini, 1997, p. 30 as cited in Bernard & Goodyear 2009, p. 149).

Relationships in supervision can be described from different point of views. It can be seen as an interaction between two or more people and/or an arrangement of different roles.

The dyadic relation of supervisor and supervisee is similar to the alliance in a therapeutic setting. The supervisor has to create a save and warm atmosphere in order to facilitate the disclosure of the supervisee. During the supervisory process the supervisee develops from a dependent position in the beginning a more independent position at the end. In the therapeutic relationship the therapist feels only responsible for his or her client. In the supervision of a trainee the supervisor is also responsible for the client of the supervisee. This can be called a triadic relationship. There exist two parallel systems: the supervisory system of the supervisor and the supervisee and the therapeutic system of the supervisee and the client. These systems are linked by the supervisee who brings information, worries and questions about the therapeutically work and relationship to the supervisory system. Going back to the therapeutic work the supervisee carries new insight, ideas and strategies he or she has got in the supervision.

In psychodrama we are used to consider the single client always connected with significant others. Psychodramatists with the concept of the 'social atom' in their mind are never working with individuals although there is may be only one client. As a therapist, counsellor or supervisor psychodramatists visualise and work unavoidable with the client's social network as well. These systems can be called parallel (in analytic terms) or isomorphic (in systemic terms) or in resonance (Elkaïm, 1995, p. 660) The situation becomes even more complex if the supervisor is working with groups of trainees who are working in groups.

Relationships in supervision can also be described by roles and their goals. Williams (1995, pp. 28-104) distinguishes between the role of the teacher, the facilitator, the consultant and the evaluator as the main roles of a supervisor. Others consider the supervisor to be a master and the supervisee to be apprentice.

The role of the *teacher* takes a bigger part especially in the beginning of supervision in training. Supervisees want their supervisor to be teachers when they start their practical work. The goals of the teachers's role is developing competencies, e.g. managing the techniques, building hypothesis based on theory, keeping structure and frame, etc.

The supervisor's role of a *facilitator* aims to create a save atmosphere and a 'holding environment' as Winnicott (1965) put it. It is "the kind of relationship that needs to be provided by supervisors so that supervisees can progressively learn to be more reflective, receptive, creative, and tolerant of ambiguity." (Geller, Farber, & Schaffer, 2010, p. 212) The supervisor as facilitator gives the supervisee also the opportunity to process feelings.

The supervisor as *consultant* collaborates with the supervisee to work out a certain problem and provides options rather than answers (Williams, 1995). The *evaluator* in contrast has to give feedback and is more responsible towards the training institute and society than to the supervisee. In some countries the supervisors as evaluators have also a gate-keeping function. For a national professional organisation they give their evaluation about the entrance of the supervisee in independent practice. Theoretical and philosophical attitudes of the supervisor and the supervisee influence which role is emphasised more.

Bateson (1977, p. 83-84) distinguishes between complementary and symmetric relationships. When supervision succeeds the relationship of the supervisor to the supervisee develops from a dependent to more independent and from a complementary to a more symmetric one during the process.

Some supervisees are in favour of a relational model as the master-apprentice relationship. This model is supported by a study of Geller, Farber, and Schaffer (2010, p. 218) who write "Indeed, without being instructed or encouraged to do so, supervisees store in long-term memory functionally useful representations of their supervisors that can serve as 'dialogic partners' when thinking through the formulations and implementations of interventions". It seems that supervisees remember strongly the role of the supervisor and refer to it when the relationship was considered to be good.

Former relational experiences with authorities like parents or teachers are influencing the relationship between supervisor and supervisee. The supervisor has to be aware that his or her way of supervising can provoke the same feelings and expectations towards the supervisor as the supervisee has had as a child.

The supervisor can use these role expectations to reflect on it and/or to provide another role model. In an example Marcia Karp[1] describes how she is "avoiding a repetitive dysfunctional attachment with supervisees."

It is important that a supervisor knows how best the supervisee can learn. Because of that, I try to tailor each supervision relationship to the specific attachment needs of the trainee. Some students like watchful guidance, and a lot of intervention. However, others like space in which to experiment. The latter needs to control their own feedback about their actions by asking questions of the supervisor rather than having the supervisor interrupt.

This has led me to ask the following questions of a potential trainee director under my supervision. As they are about to direct the selected protagonist or to conduct the group warm-up, I begin by saying:

"I would like to increase your learning as a director. I want us to have a productive working relationship. Because of this, I'd like to ask you a delicate question. What was your dominant parent like with you?"

The student director might say, "My mother was neglectful and abandoned me." I then may answer, "In that case, would you like me to be present with you and stop you when I have a suggestion. Also would you like, periodically as a double that I make suggestions as if I were you whom you can accept or reject?" Usually, the trainee feels safe and supported. Guidance from an authority figure in their lives has been missing.

Alternatively, if the potential director says," "My dominant parent was smothering, critical and I wasn't heard," I may then ask," Would you like me to give you lots of space to make decisions? I will wait for you to ask for help from me or the group." Would that be useful?

Repetition of negative dysfunctional early attachment patterns in supervision is a hot bed for unproductive learning. The trainee can feel uninspired, frustrated and endlessly criticised. These negative responses kill their love and enthusiasm for spontaneity in directing.

The contributions in this section explicate the aspects which are mentioned above. Judith Teszáry emphasises in her contribution the attitude of the supervisor towards the supervisee and explains the differences between a psychodynamic and psychodramatic approach in terms of relationship. Using a structured way of process analysis she shows her way of combining doubling, role reversal and mirroring and even giving some theoretical input in an open and discussable way where supervisees can feel free to give their ideas a go.

---

[1] Marcia Karp (2010) e-mail, 24.06.2010

Einya Artzi describes in the beginning the pros and cons of a master - apprentice – relationship and the roles of a supervisee. As a relational psychoanalyst she continues by using the theoretical concept that the psyche is formed by patterns of real and imagined relations, to introduce the reader into her model of supervision.

Sue Daniel focuses on the creation of a co-relationship in supervision. She describes the mutual development of supervisee and supervisor based on the role theory, the theory of spontaneity and sociometry. It becomes visible how the different roles of a supervisor that can be found in the main role clusters are interconnected to the development of those roles in the supervisee.

Arşaluys Kayır deals with the problem of performance anxiety within the supervision group and discusses the various reasons for it. Describing her way of being a role model and by structuring the process she is fostering self confidence and assertiveness within the supervisees step by step.

References

Bateson, G. (1972 / 1977). *Vers une écologie de l'esprit.* Paris: Seuil.

Bernard, J., & Goodyear, R. K. (2009, 4th ed.). *Fundamentals of Clinical Supervision.* Upper Saddle River (N.J.): Pearson.

Elkaïm, M. (1995). Description d'une évolution. In Elkaïm M. (Ed.) *Panorama des thérapies familiales.* (p. 587-612). Paris: Seuil.

Geller, J., Farber, B., & Schaffer, C. (2010). Representations of the supervisory dialogue and the development of psychotherapists. *Psychotherapy, Theory, Research, Practice and Training.* 47, 2, 211-220.

Williams, A. (1995). *Visual and active supervision. Roles, focus, techniques.* New York: Norton.

Winnicott, D. W. (1965). *The maturational process and the facilitating environment.* New York: International Universities Press.

The supervisory relationship in psychodrama training: More than a process

Sue Daniel

The supervisory relationship is a significant element in the development of the professional identity of a psychodramatist. It is a living relationship between two people and requires commitment, connection and mutuality. The focus of this chapter is on the relationship between the trainer and the trainee and the training journey. I define 'training' and 'supervision' and look at the three theoretical foundations of psychodrama: the spontaneity theory of learning, sociometry theory and role theory. It is vital that the trainer has a vision of what constitutes a psychodramatist, a philosophy of supervision and a contract of understanding.

1. What is training and what is supervision?

The teaching and learning of psychodrama happens through a process of training and supervision. I find it useful to delineate the difference between training and supervision although in practice the two go hand in hand. Training is more closely aligned with content, philosophy, theory, concepts and techniques, whilst supervision is concerned with professional identity and ability, which includes the trainee's adequate application of the method.

The main purpose of supervision is to enhance, develop and improve the psychodramatic skills of trainees and to assist them in their role development as psychodramatists. This enables them to increase their effectiveness in their application of the method. The relationship between the trainee and their clients is the primary focus of attention in one-to-one supervision. In the training group the relationship the trainee director makes with their protagonist is also part and parcel of supervision.

Two people are in this process of supervision, the trainer and the trainee. The relationship that they co-create is central to the development of the professional identity of the trainee. The evolution and maintenance of this relationship occurs over several years and goes through many transformations as the trainee develops. The trainer optimally grows in this process as well.

Supervision is a growth process. It is challenging and exciting and for some, it may even be a bit scary to think of entering into a relationship with another person over a long period of time. Trainees are adults who have their own unique set of abilities, life experience and values. Their social and cultural atom is already well developed. The ability of a trainer to be able to sustain intimate adult relationships is crucial to the trainee maintaining that same ability with their groups and clients. The group process also aids the development of a professional identity. This can only occur through being in an ongoing group over time, where relationships are developed with peers and trainers and by directing with supervised practice.

'Connection' is the operative word in supervision. Are we connecting with our trainees and are they connecting with us? How do we bring these connections about? I like to take a fresh look at each trainee every time we meet. I become aware of the tele, which exists between us in the first moment of meeting and at every other meeting. How is this person with me today? How am I with them? What thoughts are emerging as we talk, what is our conversation about? This warm up in myself is essential if I am to be fully involved with them. The role of 'clear seer' is therefore an elemental role for me as their trainer.

My role of sociometrist is also integral to the supervisory process. I am not standing outside the process; I am in it, knowing that I am affecting it as much as the other. It is a two-way process. The unique set of roles of trainer and trainee and the relationship between these roles comes into play, each influencing the other. Jacob Levy Moreno put forward the idea that in every single interaction with another person there is a value system (a set of personal principles and standards) involved (Moreno, 1946). The roles, which emerge in each interaction, are an expression of those values, whether we are aware or not.

2. Three theoretical foundations of psychodrama

Three interrelated theories pertain to psychodrama training and supervision, the spontaneity theory of learning, sociometry theory and role theory. Central to all three is relationship.

2.1 The spontaneity theory of learning

Inherent in a spontaneity theory of learning is the training of spontaneity states, rather than the learning of contents. The warming-up process is integral to this.

Since life presents new things constantly, consequently any session that we conduct with a person or with a group is governed by this phenomenon. To make sense out of the new through creating something out of the emergent interactions of the group is what each of us is being challenged with in our professional work as trainers and group leaders (Daniel, 1982). The more we are able to be spontaneous the more we are likely to meet the demands of any situation with novelty and adequacy. Moreno maintained that the rising and falling of the spontaneity states in a learner, effect learning. He identified two ways of learning, over learning and under learning. These are related to the theory of spontaneity, tele and warm up (Moreno, 1946). When people over learn they have no freed up spontaneity to deal with the moment, they are over rehearsed. Under learning allows a maximum of spontaneity taking into consideration the concept of the moment and a myriad of other factors in any given situation. There is more likely to be a more easy transfer of learning, a greater ease of application in a wide range of situations. Moreno maintained that "The overlearner wants a cultural conserve to stick, the underlearner wants his spontaneity to thrive." (Moreno, 1934/1978, p. 543).

Warm-up is therefore central to any activity. The warming-up process is a practical term coming from spontaneity theory. It is the working definition of spontaneity. Its purpose is to wake people up, to involve them in the moment. Numerous activities known as warm-ups and techniques bring this state into being. The protagonist, auxiliaries and group members move into motion using physical and/or mental images that lead to the attainment of the state. Techniques such as self-presentation, doubling, mirroring and role reversal facilitate this process. Thus all trainers can think about this axiom if they or their trainee gets stuck. If there are obstacles along the way, the trainer might focus on the warm-up. Sociometric questions are useful in developing the warm-up or to create new warm-ups. The trainer, working as a sociometrist, might adopt the role of the 'naive enquirer' and ask questions such as, "What is your warm-up right now?" "What is getting in the way of you being in the group right now?" "What was your warm-up coming to supervision today?" The understanding of spontaneity, known also as the 's' factor, is an important prerequisite to the learning and use of techniques and also in the application of the method. It is an integral factor in the development of vital relationships.

2.2 The theory of sociometry

The basis of the content of my training program is drawn from J.L. Moreno's theory of sociometry (Moreno, 1934), which he saw as the foundation to the methods that he later developed, to enhance, build up and stimulate the organizations, systems and communities in which we live and work. Sociometry is based on the idea of relationship and the human need to bond. Looking at the bonding nature of people rather than seeing them as fragmented and isolated is what Moreno's sociometry is all about. The human need for interdependence is inbuilt, without it we wouldn't be here. In fact, as Gavin O'Loughlin (2009) pointed out, "None of us has achieved so much as a heart beat without the support of the natural systems."

Psychodrama trainers need to be knowledgeable about sociometry and its therapeutic tele process. The tele relationship provides a means of self-correction. For example, I can find out whether my teaching and supervising methods are working adequately with each person. Both the trainee director and their auxiliaries experience and learn new ways of functioning while the other trainees might learn from the supervision of their peer. There are certain interventions or things I can say to a person later on in their training that would be unwise to do in the early days, simply because our relationship hasn't developed enough. Equally, there are things I can say early on, in order to create and deepen the relationship. The level of training, type of learner and context all influence the various interventions that any trainer might choose to use.

2.3 Role theory

The foundation of role theory is the cultural atom. The cultural atom is the pattern of role relations between a person and the significant others in their social atom. This relational theory is the embodiment of the psychodrama method. There is a genius in the twin concept of the social and cultural atom, in that it posits that we are always in relationship to someone or something at any time in our lives. Each of us affects the other. It follows that role theory is an integral part of any psychodrama training. It is necessary for all trainees to develop roles that will assist them in their ability to direct, to expand their role relationships with each other, their trainers and their significant others. Roles such as 'naive enquirer', 'spontaneous actor' and 'enjoyer of life' are essential for vital learning.

I use a role theory framework when I supervise and in my work as an individual and group psychotherapist (Daniel, 2009). The formulation into roles enables a conceptualization, which in turn enables trainers and therapists to evaluate the progress or situation of the trainee or client and their own roles and responses.

3. Having a vision in supervision

Supervision by its very name calls for the art of seeing. We can see that which is 'within', that is, our inner processes and 'without', the world around us. A supervisor, who is in the role of a 'clear seer', sees 'what is' in front of them, rather than what they think is there. When we see 'what is', we discover something. Supervision is about discovery, adventure and clarity, seeing afresh.

We want to see the growth of a director in a psychodrama training process. What are we looking for? Here is where having a 'super' vision is essential. What can we imagine we will see at the end of a process? Someone who understands what it is to be in relationship and sustain that relationship? Someone who has developed a greater perceptiveness, sensitivity and skill? Someone who has the necessary role development for adequate functioning as a director? What roles could we imagine as being essential to that image? What range of roles is available to the director? Is the director caught up in the old? Is the director working from old ideas, images, values and prejudices? Or, are they with an open mind attentive to the here and now? Our vision might be of a person who is able to be a companion on a journey, able to see, listen and be able to stand in another person's shoes. It might be of a person who has developed a range of creative expressions and who has an ever-widening ability to interact spontaneously; a person who is willing to find new creative solutions in work and play; and finally, it might be a person who appreciates the refinement of abilities and new learning. Whatever the vision, once we have one, we have a place from which to start.

4. A philosophy of supervision

As a trainer I am conscious of being in a sociometric system with others, who like me, have their own creativity, spontaneity and ideas. Part of my role as a trainer is to establish some sort of guidelines from which to work and to train and supervise people who will be practitioners of this method.

Whilst the path might be clear as to the content of the training, I am aware that the ability to learn psychodrama is sometimes quite difficult for the trainee because it involves looking at self, discovering aspects of that self that the trainee might not like, such as being seen by others, taking instruction and entering into the unknown. The establishment of a philosophy of supervision is therefore invaluable. The following philosophy may serve as an example and includes five basic principles:

- Establishing a relationship of rapport with the psychodrama trainee or practitioner with the aim of the growth of this person.
- Developing warming-up techniques as they relate to the spontaneity state of the trainee.
- Understanding the therapeutic tele process, which is at the centre of the psychodramatic process.
- Consideration of the personality makeup of the trainee, which includes knowing or finding out their limitations. What is their growing edge, what are they able to do, what are they not?
- Having knowledge of role theory (Moreno, 1946; Daniel, 2009). An understanding of role theory complements the above and vice versa.

5. The role cluster of a supervisor

Roles that a supervisor may employ are those of 'clinician', 'sociometrist' ('systems analyst', 'social investigator'), 'role theorist' and 'spontaneous actor'. They might also be a 'believer in the creative genius' and a 'creative artist'. I have identified *three primary role clusters* that make up the role of a supervisor. I specifically say 'supervisor' here because a trainer would have many more roles, such as 'training administrator' and 'curriculum developer'. Naturally there is some role overlap between the clusters because our roles are an expression of our whole selves and do not fit neatly into separate clusters:
(1) The Producer - includes roles such as 'creative artist', 'spontaneous actor', 'wise person', 'magician' and 'coach'.
(2) The Sociometrist - includes the roles of 'social analyst', which means the person can make an assessment, give a description and explain something by separating it into relevant elements. The sociometrist is also a 'social investigator', or 'explorer', who can look and see what is happening in any particular situation, at any moment of time, a 'clear seer' and 'active listener'. The sociometrist can look at the group structure continuously using the tele relations in the group as a measure and is a 'systems thinker' and 'naïve enquirer'. The sociometrist ideally enjoys discovering new perspectives and has an appreciation of life. A sociometrist is usually a 'lover of life' and is interested in relationships.
(3) The Clinician - includes the roles of 'clear seer', 'active listener', 'empathizer' and 'creative visionary'. The clinician is also a 'therapeutic guide' who is able to clarify a group and role warm up. The clinician may be a group therapist or a group leader. The clinician in psychodrama is a 'producer' and 'teacher' who knows when to apply psychodrama techniques such a maximization and

concretization. The clinician knows about conflict resolution, is a 'role model', 'coach' and 'role trainer'. They are a 'believer in the creative genius', understanding that this resides in everyone.

A psychodrama supervisor working in any field or discipline might also employ these three role clusters while maintaining the language of the culture in which they are working.

6. Supervision in the training group

The training group is the optimal context for supervision to take place. The trainer can see the functioning of the trainee director and make the necessary interventions during the drama. The supervisory relationship between me, and each of my trainees is developing from the first session. I am beginning to know them, their learning styles, how much each can take being told what to do, how much they need, and how to deliver it. They are beginning to know and to trust me. We are in a two-way learning process. I am learning too because I have to come up in myself and respond to each person. Every trainee requires a different supervision. I am cognizant of the group members who are confident but don't know much, the ones who are not confident but do know a few things, the ones who fight at the slightest suggestion, those that are anxious, passive, aggressive or doubtful. I also observe how much the trainees are learning from my directing and from watching and experiencing the supervision of their peers. When giving supervision it is important to be thoughtful in the timing, the tone of voice and the placement of any intervention. This engenders trust.

I have found that the best way to supervise a trainee is in the moment of their directing. Being 'in situ', the trainee is experiencing an emotional and sensory impact through being involved with their protagonist in the directorial situation, while experiencing full intellectual functioning. New trainees require instruction since they are learning about the method and it's techniques. As they increase in knowledge and skill they require different supervision so that they can refine their skills. The trainer has to adapt and be able to shift roles depending on the situation. They may need to employ the role of a coach, teacher or therapeutic guide. The following may serve as examples of this:

A) Coaching: *Example 1:* The trainer asks the trainee-director to tell their protagonist to reverse roles.
Example 2: The trainer asks the trainee-director to move closer or away from the protagonist, whichever is required.

Example 3: If the trainee is directing with their back to the audience, the trainer might ask the trainee/director to watch their back and keep connected with the audience.

B) Teaching: *Example 1:* The trainer asks the trainee-director to pause and express what they are thinking in a particular moment.
Example 2: The trainer asks the trainee-director to tell them what they are feeling and then asks them to breath and keep feeling as they continue to direct.
Example 3: A role training approach to directing:
If the trainee-director is stuck for what to do, the trainer might pause the drama and ask the entire group including the director and protagonist to get into pairs, discuss what is going on and then to reflect on what they might do if they were directing. The trainees then share what they discussed with the whole group. After this, the trainer tells the trainee-director to continue on with the drama. At no time is the trainee-director told what to do.

C) Therapeutic guide: *Example 1:* The trainer pauses the drama and conducts a drama themselves, this time with the trainee-director, concretizing any obstacles or roadblocks such as messages from their 'critical judge', an 'anxious Aunt' role, or even an emotional event which may have been triggered in the trainee-director while being in a parallel process with their protagonist.

Supervision involves having a definite goal or purpose. I often write a brief plan of action before I run a training or small group supervision session. I also write up sessions or make notes or role diagrams at the conclusion of a group. These role diagrams often include naming my own roles in relation to the trainees' roles. This gives me an opportunity to also consider what is going on with me and to swiftly make an evaluation. This helps my warm up to the group and the work, helps me to think clearly and make plans for subsequent sessions.

And finally, the ability to play is also an important capacity for trainers to have so that everyone enjoys their learning and has fun. We learn best when we enjoy what we do.

7. Conceptualization of the method

Working out to what degree any trainee director has comprehended and absorbed the concepts that are being taught is part of the role of a trainer. They can discern whether concepts are being learnt and integrated through observing directorial or auxiliary work and how the group members relate to each other in the group.

Giving the group members certain tasks provides a further means for discernment. For example, at the conclusion of a drama I might say to the group that I want two or three of them in turn, to come out the front and make a presentation, about where they saw role reversal happening and the result of the use of that technique. By having the trainees present to the group what they have seen, allows me to see whether the concepts I am teaching are getting across to them. Presentations may include an observation of what happened, where a certain role emerged, or the result of the application of a technique. This kind of activity takes a few minutes and as well as being diagnostic it provides an opportunity, to extend the trainees' thinking and for them to be an authority in themselves. It assists them to function independently; it focuses the group on the training aspect of the work; and it assists the trainer to know the learning style of the trainees. The trainees are also learning to be group leaders by standing in front of a group and expressing what they have seen and what they think about this.

Throughout the training process I create tasks whereby trainees can experience their own level of functioning. This assists the trainee to develop confidence in their learning abilities. Trainees in my institute direct each other in vignettes in the first session in the first year of their training. They usually produce small pieces of work and therefore we are teaching the role of 'producer' rather than the more formidable role of 'director'. They gain confidence by getting up and directing one another in these brief vignettes and by taking auxiliary roles. The acquisition of confidence allows trainees to function better in the group and increases their spontaneity. The result is a corresponding decrease in anxiety and a heightened ability to retain knowledge and to learn. The trainees understand that they are not able to go out and do full psychodramas with clients, of course, but those who are already experienced group leaders often start and use aspects of the method early on in their training. When this occurs they enter into one-to-one supervision and share with their group examples of how they are using the method. One of our trainees was working in the department of psychiatry in a large public hospital when she began her training. Half way through her second year, in consultation with her primary trainer, she started a psychodrama group for patients on a locked ward. She undertook regular supervision with a trainer/supervisor and presented aspects of her work to her training group. Nine years later, her psychodrama groups continue to be popular with staff and patients.

As the spontaneity of the trainees develops, the training group becomes a work group. Trainees focus more readily on tasks, generate ideas from their own experience, their role repertoires develop, there is a greater expansion of roles and they become more willing to display themselves. Their relationships rapidly develop, with themselves, each other and their trainer. With the development of this open group climate, the group members find themselves more easily able to analyse roles, draw accurate role diagrams and make interventions. They are able to work with increasing spontaneity as directors, auxiliaries and group members. I provide a range of activities, creating anew, thus the supervisory relationship has the opportunity to remain dynamic and adequate. The roles of the trainee director are observed and their growing edges are noted. I make interventions on this basis. I work out which role or roles a trainee needs to develop in order to progress and what interventions might be required. This is often done in consultation with the trainee, sometimes in the group and we agree on a course of action.

8. The active generation of ideas

Trainees learn ideas through the experiential method. When a trainee generates his or her own plans in relation to their peers their sociometric status increases. The following four examples illustrate this:

- Two trainees in a second year training group wanted to get more skill in learning about the technique of concretization. When they brought this forward in the training group, other group members began to think about this concept. I watched the relationships in the group develop as they discussed their thoughts. Seeing the extreme interest I decided to spend their next full day of training focusing on this technique.
- On the day, one of the trainees said she wanted to be a protagonist and to experience concretization. By focusing mainly on concretization the auxiliaries expanded their roles with the result that there was a greater spontaneity in the group and the trainee was able to go deeper into her experience.
- Another trainee, in his final year of training, suddenly realized how sociometry informed his work as a director of an organization. He had been applying sociometry very effectively in his work-groups, in staff meetings and around the Board table but had not been able to articulate *what* he did in psychodramatic and sociometric terms. This new realization prompted much celebration and a long conversation and discussion ensued in the training group.
- An advanced trainee runs groups for families with teenagers who express themselves violently. She is working with her ideas on warming up the group and using small pieces of work to create safety in the group. Whilst this is not a

psychodrama group as such, her group work includes the use of the technique of mirroring and she is working from her knowledge of sociometry.

9. Working with situations in a trainee´s work life

As the training proceeds the trainees bring to their group situations that arise in the course of their work. The following is an example of this: A trainee reported that she was having trouble trusting her immediate responses as she directed. A scene was set with the trainee sitting in her chair. The supervisor asked how her doubt manifested.

Trainee: It's like a cloud above me.
Supervisor: Come out of that chair and be the cloud (she goes behind the chair).
Supervisor to the cloud: (Interviewing for role) What are you doing to her?
Trainee as the cloud: I am making her doubt herself.
Supervisor: Let her know how you do that. (She does this.)
Supervisor: (Interviewing for role.) What else do you tell her? Tell her now.
Cloud: You need to be careful.
Trainee as herself, sitting on the chair: (To supervisor.) I feel split.
Supervisor: Get that chair over there and put it next to the other one.
Trainee in new chair: (She takes up a new role) I think what you are doing is fine. (She has a dialogue with this new aspect and discovers that she needs yet another role. This time it is the role of a mother but not her real mother.)
Trainee as the new mother: (Moving closer) You are doing well.
Trainee as herself: Thank you (She is visibly moved).

There is a clear movement toward the mother as this supervisee takes in her words and feelings. It is a beautiful moment. The relationship is one of unconditional loving companionship. The cloud has no influence over her anymore. Vignettes such as this are very useful small pieces of work that assist in the formation and integration of new roles. This drama assisted the trainee in the continuing development of an inner trusting and loving role relationship with herself. While she was in the Mother role, she was developing the ability to be a 'loving Mother' to herself. She had not had that experience in life with her own Mother, had often felt lost and unloved and in her teenage years had been addicted to drugs.

A further example is as follows:

A group of practitioners had been meeting monthly. In one session one of the members mentioned that she was very anxious about a group she was running. She had not slept for several nights, had been up early the previous day and was cleaning the house. She set a scene of her kitchen. She was washing up at 2 a.m. I asked her to walk around the room and make a soliloquy, "Speak out loud and tell the universe everything going on in you, thoughts, feelings and sensations." She did this and after a couple of minutes suddenly stopped and cried. I moved closer to her and asked her to breathe in and out several times. After the sobbing had abated, she said she realized how scared she was. The fact that she realized this through being in touch with her fear was a pivotal moment. The catharsis of abreaction enabled her anxiety to drop away and she became spontaneous. Prior to this her anxiety had been too high to let anything or anyone in or even to think clearly. She immediately became open, was able to focus on her plan and to create activities from her ideas.

The session then went for another half an hour and during this time she was able to independently continue with her ideas and make plans on the whiteboard without the debilitating anxiety. The group gained from this and each member made notes on the planning of sessions. A vibrant discussion followed.

With newer trainees however, the idea of developing roles, exploring the earlier social atom or developing an understanding of techniques is foremost in my mind. With advanced trainees or practitioners, the work involves more of a refinement and integration of roles. However, regardless of the level of training my aim is to meet the person in the moment.

10. Reading and writing in relationship

Writing in relationship is quite different to writing in isolation. It assists in the development of clear thinking, furthers conceptual abilities and brings about a further integration of feeling and thought.

Trainees are encouraged to produce small pieces of writing, a sentence or two, sometimes a role diagram or a piece of creative writing during their first year. In the second year of training the trainees write down or draw the roles they observe in the dramas and we process these using a role theory framework (Daniel, 2009). Writing is an integral part of the overall training. Every trainee has supervision on their written work and here they are often faced with issues of exposure and old fears from the past, of rejection and not being good enough. The relationship with the trainer is therefore very important. Both are tested in this process.

When the advanced trainees complete their projects, papers, theses or journal articles we put them into our library. The trainees come to appreciate that they are adding to the development of others, peers and practitioners, which adds to their self-esteem.

Having a library resource in a training institute is an excellent resource. Trainees can develop their knowledge by making use of a wide body of literature on psychodrama that relates to the ideas being taught. Whilst they borrow what interests them, their trainers also recommend articles that are relevant to each trainee, depending on their area of work, or highlights particular conceptual areas. This assists them in the formation of a positive professional identity through the idea that they are *in relationship* to something larger than themselves. They become familiar and knowledgeable not only about other trainees' work but also practitioners who live and work in other countries, their respective ideas, work and on the application of psychodrama in specific areas.

11. The professional identity of the psychodramatist

Trainees will go through different kinds of identification processes, according to different personality needs and different ways of learning. Moreno identified two ways of learning, over learning and under learning. These as I've already mentioned in this chapter are related to the theory of spontaneity, tele and warm up (Moreno, 1946). The core goal is the training of spontaneity states. This requires an understanding of spontaneity on the part of the trainer. An example of this is as follows:

Jasmine (not the real name or necessarily the gender of the trainee) was directing a psychodrama. It became clear that the protagonist would benefit from some role training. They did not have a certain role needed for the development of their relationship with their spouse. The role did not exist in the family culture either. Jasmine struggled with the role training. She lost her place, picked it up again, floundered and then picked up the reins again. She took the coaching quite well. Later though when we were processing the session, she burst into tears saying she knew nothing and it was clear she felt embarrassed about this in front of her peers. Jasmine hadn't done any role training, she had seen it done, was part of the process when others had been doing it, perhaps she had been away, she had earlier had a term away from the formal training, many things could have been the case. However what emerged was that her normal way of directing had been to follow certain guidelines and she had learnt the content of the process of the method very well. In this session she was thrown into an unknown situation unexpectedly. The processing of the session allowed the trainer to understand that Jasmine needed more spontaneity tests to further increase her ability to be spontaneous and at the next training session she enjoyed these challenges as did her peers. The trainer didn't focus on Jasmine but instead spontaneously created a training session with a range of new activities beginning with a sociodrama. A little bit of performance anxiety in the group culture dropped away and people became more relaxed and open.

Ekstein, and Wallerstein (1972), identified several kinds of learners; open learners, anxious learners and fighting learners. Knowing about different learning styles can assist trainers to appreciate their trainees' uniqueness and respect their learning styles. They can honour the uniqueness of the trainees and at the same time, work for the benefit of the trainees' and the group's progress. In this way they demonstrate that they are open to learning about each trainee and to teaching from a systemic approach.

A trainer doesn't know everything and does not need to know everything. In fact, sometimes it is a very good thing that they don't and face the trainee in the moment, staying open to the new. By doing this, the trainer is providing the trainee with a two-way learning situation, each being able to encounter the other. The trainer can relax, knowing that they are contributing to the formation of the trainee's identity. The trainer is a 'role model' for the trainee.

A professional identity is an amalgamation of two parts. Firstly it is how a person sees himself in their profession, which includes a reflection of their inner experiences, professional relationships and to some degree their life experience. This makes up an 'internal' identification process. The second part involves an 'external' identification process, which distinguishes the person as belonging to that profession. This can be measured. For example, a trainee may write a psychodrama paper, which is subsequently assessed and then passed by a board. The trainee has now met part of the requirements necessary to become a psychodramatist and the community recognizes this. The internal process, on the other hand involves a process of identification with the trainer. This is a slightly more complicated arena due to the subtleties of the relationship and the dynamics within this. A twofold process of identification takes place with the trainer: (1) As a person who has similar or different values to which the trainee will aim. (2) As a 'model' of a method and a function.

The professional identity becomes part of the person's life. The process of training in psychodrama must therefore provide an opportunity for professional self-realization and individuation. If this does not happen, then there are options, which may include the trainee finding new pathways for growth and fulfilment, or the trainer might have to work toward developing new perspectives and new roles. The relationship that the trainee and trainer co-create is central to the development of the professional identity of the trainee. The development and maintenance of this relationship occurs over a long period of time and goes through many transformations as the trainee develops. The trainer optimally grows in this process as well. It is a shared journey. How the two people grow in their relationship is an enormously creative process and rather amazing. Each are changed in and through this relationship.

Complementary and symmetrical role relationships: A formal mutually positive supervisory relationship is important for people to create some good things together. In the supervision arena, the supervisor must be a supervisor, and the supervisee, a supervisee. That is, the student must remember that they are a trainee and as such may benefit if they take in the instructions of the supervisor. The formal roles of supervisor and supervisee are complementary. What does this imply? Each has a designated role. Problems can emerge when this formal relationship is not understood by one or both. Like parent and child or night and day, complementarity involves difference. The social role relationship of a teacher and student may be described as complementary, the teacher is teaching and the student is learning. Social relationships such as, supervisor to supervisor, parent to parent and child to child can be described as symmetrical. The essence of a symmetrical role relationship is similarity.

Within any 'social' complementary or symmetrical role relationship, people bring a repertoire of roles, for example, psychodramatic (psychological) or somatic (physiological) roles, which at any point in time may be described as complementary or symmetrical. A child may be a loving companion to his father, together lovingly making a model plane, each warmed up to the other's actions, passing pieces of ply wood to each other, respectively holding and gluing, being enjoyers of life together. In that moment they are in a symmetrical role relationship since both are enjoyers of life and warm companions. The concepts good/bad, weak/strong do not apply to either complementary or symmetrical role relationships. It depends on the context and the people in a particular context. Gregory Bateson (Watzlawick, Beavin, & Jackson, 1967) coined the concepts of complementary and symmetrical. I found these terms extremely valuable for working with people and have assimilated them into my training and practice (Daniel, 2009). What is important to consider is the relationship, not whether we are right or wrong. In any relationship where there are two people, there are three entities, person a), entity 1, person b), entity 2 and their relationship c), entity 3. This concept is also very significant and useful in couple counselling and family therapy.

The ability of a trainer and trainee to be flexible and to be able to take a complementary or symmetrical role to each other requires commitment and a good measure of spontaneity and creativity. Sometimes the trainer has to be willing to be the one to make the shift. To hold such a philosophy may engender growth toward wisdom, individuality and a valuable alliance over a lifetime.

12. The development of a contract

When making a contract in individual supervision there are three main things to consider:

- The ongoing professional development of a person;
- Their reason for coming to supervision, and
- The establishment of a focus, which aids spontaneity and creates safety.

In a training group, the aim is for trainee directors to establish a mutual relationship with their protagonist through making a contract. Examples of this might be as follows: “What do you want to focus on in this drama?” “What is your aim in doing this work? OK, we are going to explore what you might say to your wife. The idea is to revisit this scene and complete something. Is this right?” These are the kinds of questions that may be asked as part of the interview phase. The making of a contract provides a direction and serves to create safety for both director and protagonist.

A supervisor might also make a contract with the trainee director at the beginning of a session by asking them what kind of supervision they would like. This might be done when the person is in their second or third year of training. Before that, much of the supervision is about learning the method and the supervisor may often interrupt the process without necessarily telling the trainee. Also new trainees might not know what kind of supervision they need. Much is learnt in the *here and now* when the person is in the process of directing. Later on the feeling of the moment is forgotten and then the supervision might become an intellectualization process rather than an experiential one. The *'here and now'* relational process brings about a greater learning in the trainee and the trainer. Which brings me to the subject of ethics.

13. Ethics

Ethics is a system of moral principles. It is all about the quality of relationship, what we do with each other, how we think and feel and act in the moment and how we get on with each other in our day-to-day interactions.

How we are in relationship with each other is dependent upon the context, the role relationship, the person, the contract and the time. What is adequate role behavior in one era may not be in another. What is acceptable in one context is not in another. The contract is relevant: If a person has come for training they need to receive training. If a person has come for supervision then they need to receive supervision. In training and supervision it is about how we are with each other and with our protagonists and clients. A contract ensures a framework from which to work and a way of behaving within that.

A code of ethics protects the trainee, the public and the professional group. It provides a reference and a framework. It gives rise to thought about two-way relationships between people and groups. To understand some basic principles of codes of practice is part and parcel of a supervision process. I find that reading and discussing a variety of material on ethics a useful activity in one-to-one supervision and also as an activity in training groups. I particularly recommend, Zerka Moreno's, "Notes on Psychodrama, Sociometry, Individual Psychotherapy and the Quest for "Unconditional Love" (Moreno, Z. T, 1972; 2006) because it addresses particular things trainees experience during the training process. This lucid work addresses the relationship between therapists and group leaders and their clients. Having such guidelines and material available is useful for trainees working in any field where they are applying the psychodramatic method.

14. The end of a journey

As the trainees become more proficient in their knowledge and understanding of psychodrama, they have developed a range of roles and an ability to direct a group. Now they are coming to an end of a process. They have also completed a project or done some writing, either a journal article or a thesis. Each is getting ready to do a practicum, which is the final step in formal training. It is not the final step in the learning process because this is a life journey and involves ongoing professional development and the sustenance of relationship with self and others.

Around this time, I ask the supervisees to write an essay, two or three pages, on their training journey. This paper includes things like when and how they began training, moments of interest, highlights, insights, difficulties or obstacles, significant people or events of significance. The activity of writing about their training journey in a few pages warms them up to their *feelings* about their training journey and empowers them; they connect more deeply with themselves and there is a greater consciousness and appreciation of themselves and others. They appreciate the fullness of their journey, the time taken and the immense 'traveling' undertaken. Sometimes they find themselves reflecting on what they have achieved, their relationships with their trainers and peers and on the method itself. This process of writing serves to deepen relationship with self and others and takes them further forward on their journey to becoming an accredited psychodramatist.

15. Conclusion

The establishment of a professional identity in the trainee is the aim of supervision. This is brought about through the relationship between the trainer and the trainee, mixed with an understanding of the philosophy of the method and the skill and ability to adequately practice it. Supervision is a delicate balance of learning, being sensitive to others and respecting relationship. A trainer requires a wide role repertoire; at times they may be provocative, at other times, firm or kind. They need to have an understanding of what it means to be in relationship and an ability to commit to a process and to the people they are training and supervising. This is a necessary requirement to ensure that their trainees learn to trust the process, learn about relationships and develop a love of the method.

References

Daniel, S. (1982). *Building a healthy group culture: A psychodramatic intervention.* Unpublished Monograph. The Zerka Moreno Library, Psychodrama Institute of Melbourne, Australia.

Daniel, S. (2009). Psychodrama, role theory and the cultural atom: New developments in role theory. In C. Baim, J. Burmeister, & M. Maciel (Eds.), *Psychodrama: Advances in theory and practice* (pp. 67-81). London & New York: Routledge.

Ekstein, R., & Wallerstein, R. S. (1958/1972). *The teaching and learning of psychotherapy* (2nd ed.). New York: The International Universities Press.

Moreno, J. L. (1934/1978). *Who shall survive?* (3rd ed.). Beacon, N.Y: Beacon House Inc.

Moreno, J. L. (1946/1985). *Psychodrama, Vol. 1.* (7th ed.). Ambler, PA: Beacon House, Inc.

Moreno, Z. T. (1972). Note on psychodrama, sociometry, individual psychotherapy and the quest for "Unconditional Love". *Group Psychotherapy and Psychodrama, 4,* 155-157.

O'Loughlin, G. (2009). In private conversation. Psychodrama Institute of Melbourne, Australia

Watzlawick, P., Beavin, J. H., & Jackson, D. (1967). *Pragmatics of human communication: A study of interactional patterns, pathologies and paradoxes*. New York: W. W. Norton & Company Inc., pp. 67-71.

Supervision – a triangle of drama in transition

Judith Teszáry

In this chapter, I present my view of the supervising process and share my way of doing supervision in different settings, such as direct supervision of the trainees in advanced training groups; supervision of assistants in my therapy group; and supervision of a supervisee's own practice. My purpose is to inspire colleagues to a greater freedom in both using their creativity and inventing new techniques. The following quotation from Søren Kierkegaard encapsulates my vision of supervision:

On the Art of Helping Others to Understand

If one is truly to succeed in leading a person to a specific place, one must first and foremost take care to find him where he is and begin there. This is the secret in the entire art of helping. Anyone who cannot do this is himself under a delusion if he thinks he is able to help someone else. In order truly to help someone else, I must understand more than he – but certainly first and foremost understand what he understands.

If I do not do that, then my greater understanding does not help him at all. If I nevertheless want to assert my greater understanding, then it is because I am vain or proud, then basically instead of benefiting him I really want to be admired by him. But all true helping begins with a humbling.

The helper must first humble himself under the person he wants to help and thereby understand that to help is not to dominate but to serve, that to help is not to be the most dominating but the most patient, that to help is a willingness for the time being to put up with being in the wrong and not understanding what the other understands (Kierkegaard, 1859).

1. Introduction

The method of supervision, in my opinion, should be congruent with the method of the therapeutic practice. Psychodrama is one of the few methods where the therapist's work is visible for the whole group, as opposed to psychoanalysis where a third person is never present in the analytic room. In family therapy, it is also quite usual for the supervisor or even a whole team to be sitting in the room reflecting (Andersen, 1991) from time to time on what is happening in the session.

To bring cases to supervision involving work with transference and counter-transference is a psychoanalytic tradition, which is the essence of the method. It does not mean that transference and counter-transference do not occur in the psychodramatic sessions, but the tool to work with them is different. We usually bring the 'original' person to the scene and do not work with the transference between the therapist and the group member as a therapeutic method. Rather, the director/therapist is 'occupied' by his/her role as director. If the transference is too strong and is an obstacle to cooperation, the session should be led by the co-therapist. Occasionally, a problem between the director and the 'original' person develops and then actually dissolves into the drama, even if the transference initially was intended for the therapist/director.

Moreover, the psychodramatist in contrast to a psychoanalyst is not neutral, but more open with her/his feelings and will even express them while reacting adequately to the situation. Her/his way of functioning corresponds better to an intersubjective view, which implies a greater involvement of the therapist in terms of spontaneity and counter transference.

I prefer, in congruence with the method, an open, visible and transparent way of supervising and prefer direct supervision when it is possible. If we consider the psychodrama work as a co-creation, it is even more important to get the protagonist's consent in what we are doing together. The director's intentions can even be spelled out loudly as an inner voice or in the form of questions.

Transference and counter-transference can be included in a supervision group, but shouldn't act as the entire focus. There are many aspects of a psychodramatic work, which are worth more attention than the transference/counter-transference question. In my supervision sessions, regardless of if they are held directly in an advanced training group or with a group of psychodramatists practicing psychodrama or in a psychodrama group with patients, I apply a model of problem-based experiential learning. Did the psychodrama production address the change the protagonist wished to achieve as well as solve problems? The focus is on the director's/supervisee's problem and not on the problem of the patient/client/protagonist.

Supervision is also a teaching process involving special cases alternating between the specific and the general. Particularly in teaching/supervising sessions, I thoroughly examine all aspects of a drama.

What skills and competencies does the supervisee need in order to manage the work? These include: creating tele with the protagonist (therapeutic alliance); exploring and finding the problem; finding the clue and choosing the consecutive scenes; producing the scenes; working through the problems; coming to a change cognitively, emotionally and physically; facilitating the group process; having psychological knowledge about the specific problem, etc. The psychodramatic process is not a linear but rather a spiral process, although there is a structure that is ruled by a certain procedure.

What was or is missing in the produced drama? Was the scene setting pure, did it help the process, or make the drama flat and undynamic? The question of the theatrical aspect of a psychodrama production for me has, if not a central role, at least a prominent one for supervision. The flow and the proper creation of the scenes help the protagonist to heighten her/his experience and promote authenticity.

Was the relation, the tele, the therapeutic alliance, between the protagonist and the director good? Was the director able to tune into the rhythm and the mood of the protagonist? The director serves as a double to the protagonist in order to come closer to his/her inner world. The purpose is to create the protagonist's world as authentically as possible.

2. Supervision with advanced trainees – some examples

The trainee director has one hour and 15 minutes to run the session. We use one-hour videotapes to record each session. This time limitation helps both the trainees to train on focusing on the essence and makes the process analyses after the drama manageable. I ask the members of the training group to be observant of the different aspects of the directing and to make notes. I divide the group thus:

- Some trainees follow the relationship between the protagonist and the director.
- Other trainees look at the scene setting from a theatrical point of view.
- Some trainees check if the auxiliaries have been properly put into the roles.
- Some trainees look at the stage management and the traffic on the scene. Were the auxiliaries left unused on the scene from an earlier scene? Has the scene been cleaned before a new one?
- Some trainees look at the exploration of the problem.

Ways of executing process analysis in my advanced training group: The trainee director does her/his process analysis first, then comes the trainees' and group members' process analysis, followed by my own. First, I point out the positive qualities as well as pose questions that I have concerning the directing. The supervisee reflects on my comments and then tries various alterations to the scenes. After the performance, I also show different ways the scene might be performed, but using stand-in performers from the group, not the original players. During this time, even the group members may have suggestions and show different solutions. It turns out that many times the protagonist watching these reparation scenes gains further benefit. Frequently, the protagonist wishes to go into the restored scene to re-experience it, now in a different way, as illustrated in the following example:

A young man loved music and loved dancing. He did it when he came home from school when the father, who did not like the son being engaged in such an activity, was not at home. The father was worried that the son might tend to become homosexual. In the original scene the young man reversed roles with the radio once and imitated some music piece, but in the remake in supervision, we[1] *put a whole orchestra, which played the music he liked. When the scene was ready, the protagonist wished to go into that scene to re-experience it — both as a part of the orchestra and as the dancer. The scene became much more vivid and the protagonist experienced how this new version of the scene could give a larger freedom to be spontaneous.*

Role reversal of the director with the protagonist: Sometimes I have an agreement with the director to intervene during the drama, like in the following example:

The director is doing the first interview with the protagonist. They go around the scene. At a certain moment I ask the director to reverse roles with the protagonist and repeat what the protagonist said. I also ask the director to orally express what kinds of feeling she/he experiences while she/he is repeating the narrative of the protagonist. This helps to create tele and the protagonist often feels better understood. If we follow Moreno's postulate – that the closest we can come to another person is through role reversal – it is quite logical that this applies also in the protagonist-director relationship. The director in the role of the protagonist can even put questions to the protagonist-as-director. The protagonist-as-director can hence come up with answers as to how to continue. This role reversal helps to empower and involve the protagonist in creating the drama. I encourage the director and protagonist to do this role reversal a couple of times until a clear picture of the protagonist's problem emerges.

Dialog between the supervisor and the director: Sometimes I even stop the drama to ask the director about her/his intentions, the thoughts behind the choice of the next scene or about the problem itself. Prejudices and own interpretations are often quite noticeable and can be immediately confirmed, declined or corrected by the protagonist. What would the benefit for the protagonist be if she/he performed a particular scene, and what could that scene lead to? How is the problem understood?

This interruption and open dialogue does not commonly disturb the protagonist. On the contrary, this dialogue helps her/him both to find clues and information, and to choose the most suitable scene or to make suggestions regarding the reworking of a scene. It even happens that while I am talking to the director, the protagonist goes through a catharsis while she/he experiences the specific scene in fantasy. We call these imagined scenes 'virtual dramas.'

Direct intervention – the supervisor is directing the director: When the climate in a training group is status-free and open for experiments, reflections and non-judgmental attitudes, one can make direct interventions. Occasionally, the supervisee needs to remind the director of what she/he might already know. Increasing the spontaneity and finding out protagonist's strengths are both examples of such a reminder. To regain the protagonist's energy and creative cooperativeness, the supervisee works to reveal the protagonist's spontaneity in scenes when she/he had both power in life and mastery of a situation. This is especially valid when the protagonist sees her/his situation as hopeless and does not believe that anything can help. An example:

A male protagonist is telling the director in the initial interview of several circumstances, both in the past and in the present, when he felt ashamed and neglected. The director asks which episode was his most painful and humiliating. The protagonist initially finds this difficult, but eventually chooses the most humiliating situation he had been through at age 12, which had a significant influence on his life. The protagonist had very low energy level and not much spontaneity. If the protagonist is in a low-level spontaneous state, it is always advisable to go to scenes that explore the protagonist's psychological healthy experiences. The director needs to go to the periphery of the trauma, rather than going directly to the centre. I asked the director to go to scenes where the protagonist had confidence. Thus the director asked if the protagonist experienced any fun moments at the age of 12.

Yes, he replied. He had one good friend. In some short scenes, the two boys play out their adventures. They smoke rolled leaves in the forest; build a crib on top of a tree. The rejuvenation of the protagonist and the playful scenes fill the protagonist, the group and the director with new energy. For the protagonist, this is quite significant as it relieves some of his hardship in dealing with a very humiliating situation.

The protagonist was then able to continue with more initiative and thereby enact the scene, which is summarised as follows: The whole class was at a swimming practice. The swimming teacher was an authoritarian, almost fascist man, and the boy didn't even dare to ask to go to the bathroom when he was in need. So he relieved himself in the water and the result was quite visible. The training had to be stopped and the whole pool drained. The whole situation was very frightening. He didn't dare admit to being the guilty one, but the teacher looked into every boy's swimming pants and denounced him. The next day at school, no one wanted to sit beside him in the classroom and his friends called him many hurtful names. At the end of the re-enacted drama, a reparative scene comes when the boy's father accompanies his son to school so the father can talk to the class to ask if the students wanted the boy to continue in the class after the pool incident. Otherwise, the father was ready to take his son out of this school. The students answered that of course they wanted him to stay and then almost competed with one another to sit beside the boy. When playing the role of the father, the protagonist both regained his self-esteem and found he could represent himself in a very convincing way.

Apart from supervision of advanced trainees in the training group, two other settings are important in my supervisory work, which will be illustrated in two further examples:

Supervision of an assistant in my therapy group – an example: A 32-year-old man still had no experience of sexual relationships with women. With his religious background and strict moral upbringing, he perceived sexuality as something dirty. The director chose to work on the man's relationship to his mother, who kept her son very close and controlled him in every way she could. The man experienced some release in confronting his mother by telling her how hurt he had felt by her negative messages about girls and sexuality, but this wasn't enough. The role as a capable male lover was so underdeveloped that something more had to be done. I asked the director to ask the man what he was passionate about. He was a certified forester and his hobby was making films about animals in the forest. He loved the huge northern deer, the moose and the elk, especially the male with the enormous flattened antlers.

For the dramatization, we re-enacted one of his films and the man played all of the different roles. In one scene, he chose women from the group to represent the trees and he was a moose. He wandered around the trees and gently scraped his back on the trees (women), sniffing them and enjoying himself. After this drama, he was asked to play varying roles of husband and boyfriend. After the group finished, in real life, he went on to eventually marry a woman.

Supervision of a supervisee's own practice – an example: The supervisee is working at a drug rehabilitation hospital, leading a psychodrama group together with a colleague. She is doing her first practice in psychodrama. Group members are hospital residents and before they join the group, an interview is done with each participant. At the latest session, however, a new male member comes to the group without having been interviewed beforehand. He only has a verbal referral from the chief psychologist. The man was initially very seductive towards my supervisee and after the session, he told her that he had fallen in love with her. He went on to explain that he had had a previous relationship with another psychologist, the prison psychologist. The supervisee felt very confused. The man was extremely good looking, with an attractive physical form, and she felt herself a bit dull and unremarkable. When she later read his medical records, she became even more confused and quite frightened by his actions resulting from his split personality. Prior to entering the hospital for drug-related rehabilitation, he had spent several years in prison for strangling a prostitute.

We reconstructed the scene of the meeting with the man. I let her see the scene in mirror and she was able to recognize her behaviour as that of a young girl on a date, smiling and feeling flattered. The auxiliaries exaggerated the gestures, facial expressions and scene in general to create a more overtly explicit scene.

I asked my supervisee to choose one group member for the role of her seduced self, another for the man's seducer self, as well as two other auxiliaries: one for her professional self and one for the dark shadowy side of the man, the murderer. I purposely separated the double messages into two roles in order to make the split between both of them tangible. I let her switch around among all of the roles involved, including the dark shadow that encourages the smiling seductive man to say: "You can get her, she is already yours."

The smiling man then says to her: "You are so beautiful. I admire you." The smiling man has another voice as well: "I wish to be a normal person, the same as everyone else. I wish to live a normal life, and I feel you can help me."

"It can really be a challenge for a young ambitious psychologist," I said.

"Yes, I believed I could help him. He seemed very honest in the group saying that," said my supervisee.

"That is the feature of a psychopath and we tend to believe what they say," I replied.

The role reversal went back and forth. The auxiliaries developed the roles further. We brought back her professional role and strengthened it. The group members were invited to show what they would do in this particular situation. The supervisee was thus able to look at it from the outside and when she then took the professional role again, she used the essence of the role models of her peers. She sent away her seduced self from the working situation, telling herself: "Go to parties or whatever, and find a boyfriend for yourself somewhere else, but not here."

In the role of the professional, she became very provoked when he said: "It is an easy match, I have done it before, I always succeed with women." She became very angry and fought him physically, which gave her even more power to be further convinced in her professional role. In a new scene, she has a dialogue with the chief psychologist about boundaries and routines, which then influences the regulation of the intake procedure. This supervision could not have been done in a direct supervision with the patient group.

Role analysis, role differentiation, role training, mirroring the situation, forum play and future scenes were used in this supervision. The theoretical bases of psychopathology, drug addiction, abuse and exploitation were also explored and integrated into the supervision process.

3. Reflexions and recommendations

We use supervision to 'see clearer'; to offer an opportunity for reflection, support, guidance, mirroring the work done, developing competencies and skills as well as a professional identity and in certain contexts, even an assessment of the supervisee.

3.1 Different aims and types of supervision

In supervision it is necessary to point out the therapeutic effects of theatre aspects: It is not obvious that our trainees, who are mostly psychologists, teachers, social workers, priests and psychiatrists, understand how a drama is built up. Theatre is a performing art and psychodrama performance should express, heighten and deepen the experience of both the protagonist and of the group. It is a vital part of the "working through" process in the drama, and has an important methodological contribution to the therapeutic effect.

It is not by accident that Moreno used the form and terminology of theatre for his method: the stage; scenes; and actors – who are the protagonists, the auxiliaries as well as the group; and the choir, or audience. In group supervision I ask the supervisee to recreate scenes from dramas and develop the artistic aspects of them.

Furthermore supervisory work should integrate reflection about a more general theoretical background of the work, e.g., how developmental theory is connected to the drama. One of my students had her first practice with older people at an elderly home. Knowledge of existential, psychological and biological problems of aging was a necessary complement as a theoretical frame of reference in the production of dramas.

Age-related aspects of the protagonist's problem are important. If the narrative of the main problem through several scenes leads to the teenage phase of the protagonist's life, we usually connect the developmental theory according to that age. What are life's main tasks to solve at that particular age, and how do they correspond or mirror the protagonist's problems? For example: identity, sexuality, separation, autonomy, teenage group influence, values, norms, ideals, etc. To be mindful about the age-related aspects helps the director to form adequate questions.

3.2 Benefits and disadvantages in different kinds of supervision

Direct supervision: I often give very concrete ideas to broaden the possibilities, but the director/supervisee is the one who effectuates these ideas. The supervisee can also ask for advice if she/he feels stuck.

The benefit of direct supervision is that the supervisee can immediately try out and practice their new vision of the matter. Learning by doing. It trains the supervisee's ability to see the process, the situation or the question with the eyes of the supervisor, while still maintaining their own view. It trains the supervisee to integrate the new vision or solution into his/her own relevant practice. Direct supervision trains flexibility in the psychodramatic process. It also trains cooperativeness. Both the protagonist and the supervisee feel safe in the presence of a more experienced psychodramatist.

The disadvantage, however, can be that the supervisee loses their own flow and relies too much on the supervisor. The protagonist might listen more to the supervisor and lose contact with the director. The supervisee might then lose their confidence and authority.

Group supervision regarding the supervisee's practice: Benefits: The supervisees are freer to talk openly and explore their own processes, feelings and thoughts about the group and the protagonists. The supervisees can explore their relationship to both the protagonist and the group and deal with their own personal resistance.

It gives an opportunity to share good practice with the peer supervisees. The supervisees have the space to reflect and act out their wildest fantasies in dramatic forms totally freely. It is a suitable context for working with countertransference matters. Strong feelings, especially negative ones, but even overstated positive feelings toward the protagonist, can be explored and worked through. The supervisee can play scenes of his/her own life connected to the topic of the protagonist's problem or personality.

Parameters regarding boundaries, structure and content of the supervision should be very well established. If it is an intimate professional group and if the group has an appropriate agreement, even therapeutic needs can be met if it contributes to a more professional functioning of the supervisee.

Disadvantages: The supervisor's understanding of the protagonist's problem, her/his understanding of the supervisee's understanding of the problem and the supervisor's own intentions can differ and diverge and get lost from the core issue. Many levels of this process have to be united and coordinated in order to be helpful.

Whatever setting or technique in supervision is used, it is important to put emphasis on the fact that not only trainee-directors need supervision for their learning, but also the supervisors themselves. As much as the trainees or the practitioners need supervision to maintain their professionalism, develop further skills and competences, supervisors also need reflection groups, method development groups, and intervision groups to which to belong.

References

Andersen, T. (1991). *Reflekterande processer.* Stockholm: Mareld.

Berglind, H. (Ed.) (1998). *Skapande ögonblick, Psykodrama och Sociodrama.* Stockholm: Bokförlaget Cura.

Boss, M. (1984). *Existential Foundations of Medicine and Psychology.* New York: Jason Aronson.

Kellermann, P. F. (1992). *Focus on psychodrama: The therapeutic aspects of psychodrama.* London: Jessica Kingsley.

Kilminster, S. M., & Jolly, B. C. (2000). Effective supervision in clinical practice settings: a literature review. *Medical Education, 34*, 827–840.

Kierkegaard, S. (1962). The point of view (1859). In *Kirkegaard's Writing,* 22, 27-30.

Tomm, K. (1989). *Systemisk intervjuteknik. En utveckling av det terapeutiska samtalet.* Stockholm: Mareld.

Williams, A. (1995). *Visual & Active Supervision, Roles, Focus, Technique.* New York and London: Norton & Company.

Relational aspects of the psychodramatic supervision social atom: supervisor – supervisee – patient

Einya Artzi

I have come to practice psychodramatic supervision in response to a demand coming from the field. After the first generations of my students had graduated and begun working, I was asked by them and by students of other teachers to offer supervision. Since then, a psychodramatic supervision format began to formulate, developing into my doctoral dissertation (Artzi, 2001) and later on into a 'catalyst' for an organization of a supervision plan at the Kibbutzim College of Education's – Expressive Therapy Training Center. My first steps as a supervisor were accompanied by three dear people, who were role models to me and a source of inspiration in my path as supervisor: J. L. Moreno, whom I have been privileged enough to know in his last year, was the one to instil in me the faith in his deep ideas of action, the encounter, spontaneity, the role theory and others. Zerka Moreno was my mentor, the enchanting psychodramatic skills and humane containment of whom I had followed. Claire Danielson[1] was my first supervisor outside the Moreno Institute at Beacon, and it was she who taught me the importance of the non-symbiotic equality psychodramatic approach in respect to patients[2]. The paths of these teachers and supervisors of mine I have adopted, and they accompany now, as then.

The present paper aims at discussing some relational aspects occurring among the participants in the psychodramatic supervision: supervisor, supervisee and patient. The paper will examine the interpersonal relations formed within psychodramatic supervision[3], from the point of view of relational psychoanalysis, which understands the psyche through patterns of real or imagined relationships. Examining the relational aspect may clarify and elucidate that

---

[1] Claire Danielson (1935-2007), Ph.D., T.E.P. Adult Educator, Mediator/Trainer. She was Moreno and Zerka Moreno's student. She had deep commitment to nonviolent conflict resolution, mediation, and 'restorative justice'.

[2] The non-symbiotic equality psychodramatic approach emphasizes the supervisor's responsibility in guarding the borders of mutuality between supervisor and supervisee.

[3] In Hebrew, the word 'Supervision (Hadracha)' means 'teaching the way, or directing in the correct way'. In other words, there is a road that one must take, while the other, according to this approach, must teach. The word supervision in other languages has a connotation of inspection and criticism. Although these qualities are part of the training process, they are not its main factor.

which occurs between supervisor and supervisee in supervision, allowing a consideration of this complex setup as a network of relations interwoven in the here and now, yet at the same time also echoing the distance 'there' of the past.

The psychodramatic praxis is complex and multilayered. On the one hand, psychodramatists are required to assimilate in the course of action the psychodramatic theoretical foundation, following the Morenian method, and on the other hand they are asked to act spontaneously and creatively out of their personal and unconscious world. The psychodramatic learning process requires, therefore, special ways of instruction, which include acquisition of knowledge, techniques and specific skills, as well as refining of the natural and personal capabilities. The aim of psychodramatic supervision is to better the work of the psychodramatists/supervisees so that they could offer the patients the most influential and effective service. The practical action, in the sense of 'the doer is the knower', and the psychodramatic supervision are supposed to encourage the supervisees to acquire their professional identity, comprised of cognitive and emotional aspects.

The psychodramatic supervision learning process gives supervisees rational anchors that help them strengthen their professional selves. They also offer them basic assumptions that aid them in understanding the development and forces of the psychodramatic therapeutic process and the 'here and now' setting.

1. Supervision and teaching psychodrama

The emphasis in supervision is on imparting theoretical and practical knowledge to supervisees (Williams, 1995). The supervisor facilitates processing, analysing his/her own as well as the supervisees' work, refers to sources, reads their papers and more. Yet the psychodramatic supervisor is also, at the same time, responsible for a different kind of knowledge, in which he/she serves as supervisor-therapist, who understands that empowering supervisees as future therapists requires a reflection upon personality – emotional issues and their processing. He/she portrays the roles of containing, holding, listening and guiding; he/she is a psychodramatist of the supervisees' personality-emotional psychodramas.

One main question is how the knowledge can be transmitted. Here, the option is twofold: the role of the supervisor as master, transmitting the knowledge to his/her apprentices by means of role modelling, or the role of educator, emphasizing the cultivation of the autonomous personalities of the supervisees.

The notion of the roles of master and apprentice is complex and holds potential dangers and so a clarification is in order. Apprenticeship is part of many training situations in the therapeutic field. It may be distinctly identified in psychodramatic training, since its theatrical qualities brings it closer to the realm of art. As we know, until the modern period, the art student was an artist's apprentice. The supervisor as master may teach the psychodramatic method, techniques, patterns of actions, axioms and norms. Yet this role holds in store also a temptation for the supervisor. He/she may regard their mastership as an expression of their unique, perhaps even magical, charisma, only strengthened with knowledge and experience. In doing so, he/she may satisfy his/her own narcissistic needs. Another potential danger is that the supervisees-apprentices will nourish this tendency due to their own need to idolize the supervisor and over-identify with him/her.

The role of educator, cultivating the autonomy of the supervisee, will allow the supervisor to keep the idea of apprenticeship in proportion. This role expresses the supervisor's understanding of the importance of supporting and strengthening the individual development of each of his/her supervisees. The supervisor aware of the importance of forsaking idealization will direct supervisees toward a strong self with personal and professional liberty and ethical integrity. This liberty is a condition for their spontaneity and creativity.

2. Essential roles of the supervisor and the supervisee

These two axes reveal an important part of the multifaceted roles of supervisor and supervisee. In a paper entitled "The Role Concept; a Bridge between Psychiatry and Sociology" (1961), Moreno writes, "Role can be defined as the actual and tangible forms, which the self takes. We thus define the role as the functioning form the individual assumes in the specific moment he reacts to a specific situation in which other persons or objects are involved... Every role has two sides, a private and a collective side".

The roles of the certified psychodramatist are at the base of the role repertoire. The three elementary roles Moreno (1972) writes about are producer, therapist and analyst. Kellerman (1992) added the important role of group leader to this list. The most essential difference between all the roles of the supervisor and those of the supervisee is that, in the relational setting between them, the supervisor is in the position of giver and the supervisee in the position of receiver. The position of giver has to do with the supervisor's authority and responsibility for the supervision framework and the supervisee in all stages of the process. The supervisor also assumes an active role of responsibility for the patient. The supervisor's role of responsibility includes the maintaining of emotions arising within himself/herself during supervision. It is his/her duty to come up with appropriate ways to avoid colouring the process with his/her own emotions but rather to benefit from them and to share them with his/her supervisee.

The roles of the supervisee in the relational setting are characterized by contradictions, reflected in the basic role position, which combines the supervisor's role of 'giver' and the supervisee's role of 'receiver'. A few clusters of roles are part of this position:

- A cluster of roles having to do with personal-emotional processes vis-a-vis cognitive practical ones. The personal-emotional development of the supervisee may entail such unsettling experiences as impotence, guilt and anxiety of failure. He/she may experience them vis-a-vis his/her patients, employers, supervisor and peer supervisees in the supervision group. At the same time, his/her professional development requires him/her to assume the role of a student dedicated to acquiring knowledge and skills.
- A cluster of his/her roles as member of the supervision group. On one hand, he/she engages in the intricacies of interpersonal and group communication. On the other hand, as a professional psychodramatist, he/she is also committed to aiding his peers as a double, as an auxiliary ego and as a provider of useful sharing.
- A cluster which includes the development of his/her professional roles by means of following his/her supervisor's model alongside the development of professional roles based on his/her unique personality.

These clusters of roles together are complex and at times burdensome and confusing. It is the supervisor's role to offer the supervisee the understanding and containment of personal and interpersonal difficulties and at the same time to impart the knowledge essential for a professional.

3. Relational aspects between the roles of supervisor and supervisee

'Relationality' is a key concept in the Relational school of psychoanalysis. It suggests that the encounter between two people here and now is mutual and so are the interpersonal relationships formed early in a person's life. This encounter emphasizes the role of the part of the real or imagined relations in the dialogue. Rolef underlines that "relational psychoanalysis considers the psyche as being formed out of relationship patterns" (2010, p. 3). While these ideas describe relational psychoanalysis, they are close to the basic psychodramatic ideas. There is one concept, which may accurately describe what goes on between supervisor and supervisee in the course of psychodramatic supervision.

Psychodramatic 'supervision social atom'[4]*:* The 'psychodramatic supervision social atom' presents the relational supervision setting: supervisee-supervisor-patient in the clearest manner. As such, the supervision social atom may represent for the supervisee significant others who appear in his personal social atom, corresponding to his telic relationships. Moreno writes:

> The social atom is the nucleus of all individuals toward who a person is emotionally related or who are related to him at the same time. It is the smallest nucleus of an emotionally toned inter-personal pattern in the social universe. The social atom reaches as far as one's tele reaches other persons. It is therefore also called the tele range of an individual. It has an important operational function in the formation of a society (Moreno, 1972, p. 184).

In supervision, the relations between the supervisor, supervisee and patient create a unique social atom. In the space of this social atom a real dimension exists, in which the supervisor and supervisee are present as well as a 'transferred' dimension of the patient, which will be illustrated in the following example:

Pam, is a third year student, and participates in my weekly supervision group of eight participants in a psychodrama training school. Like her peers, Pam directs alone and for the first time, a psychodrama group of ten participants aged 30-50 diagnosed with learning disabilities and adjustment and functioning difficulties, some with minor brain damage. In the course of the supervision session, Pam describes her difficulty accepting Mira, one of the patients in the group, as well as anger, disquiet and also guilt she feels toward her. She tells about her last session with her patients:

Pam enters the room, where she observes a conflict between two group members. Mira is yelling at Tali, because she feels insulted. Pam is already disquieted and starts the session with mixed feelings:

First Scene: The patients share feelings. Mira says that she is being insulted in the group and that this is not what she came to the group for. Pam encourages a short clarification between Mira and Tali, which calms them both down, only supposedly, so it will later turn out. Yet her growing discontentment begets disorientation, leaving her unfocused to the extent that disturbs her work.

Second Scene: Kobi – another group member – is presenting Peter Pan in a vignette, who sprinkles fairy powder over the children, played by all partici-

[4] In this chapter I use the concept of the 'social atom' in a unique way in order to elaborate the relational dynamics in supervision. In this case the term 'social atom' does not refer to the social nucleus of one individual, but to the social relationship of supervisor-supervisee-patient.

pants in the group. Mira starts laughing and Nomi scorns her, saying it is inappropriate to laugh at that moment. Mira, feeling very hurt, asks what she has done, leaves the stage and declares that she wants to go home. The work gets stuck and the feelings blocked. Pam experiences a vague mixture of anger at Mira due to her lack of cooperation, self-criticism and guilt. She convinces Mira to stay and asks everyone to sit down. In this moment she is unspontaneous and lethargic, and she can hardly give a supportive feedback to Kobi.

These scenes exemplify professional issues that demonstrate the natural lack of skill and experience of a beginning psychodramatist, such as how to clarify a conflict between group members by using psychodramatic techniques or by asking Mira to be the protagonist. In addition, Pam's personal-emotional issues come up, such as anger, insult and guilt. They are catalysed by the patient's behaviour, causing unfinished business with her, confusion and helplessness, augmenting her insufficient professional conduct.

The supervisor suggests, that Pam should reverse role with Mira, because it has the potential to offer her an understanding and to resume the relation with Mira. In addition, it may allow Pam to feel relief in the places in which she unconsciously identifies with Mira's anger, insult and wish to flee.

At the beginning it is difficult for Pam to reverse role with Mira and to get her feelings, but later on Pam could understand better Mira's real or imagined insults. With the help of the other group members, the scene of Kobi's vignette is reconstructed, in which Mira started to laugh and received criticism by another group member:

Nomi: Stop laughing.
Mira (still laughing, and with ostensible acceptance): ok, ok, so I won't laugh.
Pam: What's going on?
Mira: (leaves the stage, angry) I was already insulted twice, I am leaving.

The work returns to the point of blocking. Pam softly addresses Kobi and the participants sit down.

After having these scene reconstructed, the supervisor proposes a special form of role training, in which participants in the supervision group can take the role of Pam as a director. One after the other Anat, Jan, and Ben reverse role with Pam and speak to Mira:

Anat: You may choose to leave.
Mira: No, I'm leaving.
Jan: If you don't want to, you can leave and come again next time.

Mira: I am leaving and I won't talk to you anymore.
Ben: Mira, we are about to finish working now, so I will be available to talk.
Mira: Maybe, maybe.
Anat: (changes position and stands as a double next to and behind 'Mira') I was so insulted, for the second time; so insulted.
Mira: Yes... yes...
Ben: (in front of her) Maybe it's ok for you Kobi to sprinkle some fairy powder over Mira and give her power? (Kobi agrees).
Mira: Yes... yes... I screwed up and I have someone to console me... it connects me to another place, maybe safer... to strength.
Supervisor: And you do have strength?
Mira: Yes. Yes.
Supervisor: And it's a good thing.

In the course of the sharing, Pam says that the last things Anat and Jan said allowed her to feel differently. In the processing, a week later, Pam adds that the work and especially the role reversal with Mira drew her nearer to her and to her constant feeling of insult, whether just or unjust. She felt that the idea of immediate double, partaking in the occurrences on stage, echoing Mira's insult, opened a new door for her to accept Mira.

This example of a supervision session illustrates the 'supervision social atom', which refers to the fabric of relations between the supervisor, the supervisee and the patient. While at the same time it is important to remember that this social atom is part of a cluster of other social atoms, two important ones among which are clearly reflected in the role play.

The first, the patient's social atom, delineating Pam's relations with the members of the group, especially with Kobi, with whom she immediately identified, and with Mira, toward which she had felt an immediate aversion. The second is an 'emotional social atom' (Hollander 1974) of characters from her internal world, which were not yet presented, nor was the pain that has to do with them.

In the course of Pam's supervision, it is impossible to ignore the personal-emotional issues. When the insult and flight came up and were processed, Pam agreed that they had struck the chords of her own personal pain. To such an extent that in the vignette, she 'took Kobi's side' and denied the places in which she identified with Mira.

It is no wonder, then, that in that session she grew anxious, her spontaneity decreased and she could not find a psychodramatic intervention appropriate to the situation. The tension between the professional roles that Pam had to portray and the emotional roles in her inner world is a crucial factor in the conflictual relational setting.

As a first year psychodrama director, the goal-oriented role playing brought Pam closer to her strengths and allowed her to be in a more stable place vis-a-vis Mira. The effective role playing gave Pam some basic ideas for a dialogue with Mira, which she could develop, and also brought with it some relief.

In her second year of studies, Pam was a student of mine in psychodrama. During that time I came to recognize her expressions of pain which occasionally come up during psychodramatic or sociometric processes. In these instances she 'escaped' into herself, physically shrinking herself, making herself as minimal as possible. This usually occurred in places where she experienced hurt and lack of trust in the group and me.

From then on, Pam went through a series of meaningful and deep developmental levels which allowed her to come into closer contact with her inner strengths. Likewise, I learned to understand her unique language through which she was asking for intimacy and support from me, while signalling for me to do all that at a slower pace that took into consideration her feelings of invasiveness. Together, we created a system of mutual trust that allowed us a constructive and creative 'psychodramatic trialogue'.

The 'supervision trialogue': In the supervision social atom, a combination of dialogues between real participants is created: supervisor-supervisee and the supervisee and his/her transferred participant: the patient. More accurately, a 'trialogue' (Kron, 1995, p. 183) is created. The idea of dialogue is related to the 'encounter' and at times identified with it. It is a very basic psychodramatic idea. It may also be applied to the three participants of the trialogue. Moreno (1973) writes: "Meeting means more than a vague interpersonal relation, it means that two or more actors meet, not only to face one another, but to live and experience each other, as actors each in his own right" (as cited in Marineau, 1989, p. 150).

Even though in the trialogue the relational setting discourse is not a 'face to face' encounter in its entirety among its participants, it is guided by the basic principle of seeing the other, like in a dialogue. Although the supervisor is the deciding authority of the supervision process, the case is not of inequality between supervisor and supervisee or between these two and the patient. Rather, it is a trialogue, recognizing the borders of the relations, marked by a hierarchy of knowledge and experience and an affirmation establishing the recognition of the otherness of each one.

The trialogue creates real and transferred mutual relational settings, highly affecting all of its participants. The supervisor is therefore expected to sensitively guard the borders of mutuality, making sure that symbiosis is not formed among the participants.

4. Role playing in supervision

In supervision it is important to offer a goal-oriented professional role training. The themes that came up in supervision can be processed by different forms of role play. As a consequence, the role plays used are goal-oriented role plays, behavioural role-play, and problem solving role-play. In addition, in supervision there is room for what will be called here classic therapeutic psychodrama as the core of the process or as part of it (Kipper, 1967). If there is a need of the supervisee, supervision should provide time and space for the processing of distress, which results from past encounters and which affects the supervisee's 'here and now'. The following example illustrates a goal-oriented role play combined with classic therapeutic psychodramatic elements:

Sali was my student a few years ago and now is part of my private psychodrama supervision group for psychodramatists with at least two years of experience. Sali has six years of experience as a psychodramatist. In supervision she speaks of her growing sense of anxiety concerning an ambiguity to her position in her workplace. The distress and stress there affect her psychodramatic work. She sees the patients as if through a screen and she is somewhat robotic, lacking spontaneity. The unfinished business with her employer, having to do with her job description and territory, leave her fluctuating between gratitude and admiration of her employer, who picked her up from the school bench, and feeling worthless and stuck in comparison to her.

First scene: A conversation she didn't have with her employer. These ambivalent feelings emerge. After a few minutes of reluctance, she reverses roles with her, stands on a chair and says things to the following effect: "My creativity is so special, I am both kind and talented, me, me, me..." Sali starts enjoying the character, choosing dolled up words, using exaggerated hand gestures, creating a grandiose character. Here she stops for a minute and says that somehow these gestures are hers, because she has a tendency to show off, exaggerate, being prone to overdramatic presence. She experiences fear, because it touches upon her old anxieties.

Second scene: Recreation. Sali speaks of a memory from the age of seven, in which her mother, three year old sister and father participate. She focuses on a moment in which she feels distressed. Her father, who until that point held her in his arms, deserted her, and chose her sister over her, again and again, as his special child. From that point, nothing that she does satisfies him. When she reverses roles with him her angers grows.

Third scene: A conversation that has not occurred. Sali is angry with the father yet also with herself for still trying to resemble him and his 'bombastic' and all-knowing presence- in particular. She is also angry with her efforts to please him. She intensely and painfully expresses her anger. 'The father' ignores her at first, his answers are manipulative, yet the more 'he' listens, 'he' is willing to understand that 'he' had hurt her. The dialogue takes on a different tone, somewhat settled. 'The father' whispers that he loves her and asks for forgiveness.

Fourth scene: A conversation that has not occurred. Sali invites her employer and also her private psychodramatic supervisor. 'The employer' is no longer on a chair and assumes the role of mostly listening. 'The supervisor' declares her support, patience and faith in Sali's capabilities and professional process of growth. Sali listens to her words of support with some satisfaction, yet sits a bit far from her, saying that although she needs support from both of them, her growth requires that she learns what she needs in general and what she needs to adopt from each of them. After the work was done, Sali was contemplative, and in the processing she shared a feeling of contemplation over the material that came up and said that in addition she feels that some load is off her shoulders.

In the processing in the next meeting, Sali said that she had spoken to her employer. Her duties and territory at work were reorganized and she felt more at ease, having expressed her needs to her employer. In addition, she felt that as a psychodramatist she is more free and available for her patients, and that the issues that emerged at the role play touch upon her inner insecurity, drawing her nearer to a deep past distress.

The first steps toward improving Sali's work entailed a processing of the difficulty vis-a-vis her employer and an encounter with sensitive emotional issues. To achieve these goals, a role play combining professional and personal needs was chosen. A relational setting that emerged in the role play presented the relation between Sali's dialogue with her employer and her dialogue with her father, allowing her better communication with her patients. This work allowed her to grow stronger as a professional psychodramatist.

5. Summary

In supervision, the supervisor and supervisee have a shared goal of offering patients the best psychodramatic processes possible. For the supervisee, the goal of supervision is to create a dialogue that will beget a change with him/her that will aid him/her in becoming a valuable professional.

The relational setting between supervisor and supervisee occupies a central role in supervision, expressed in the encounter between their different roles. This setting is translated into a unique supervision social atom, in which the supervisor and supervisee have a real presence, while the patients' presence is transferred. Together they created an extended version of the human dialogue – the trialogue: supervisor, supervisee and patient. That is a place in which the recognition of otherness is important, allowing a containment of the hierarchical differences of knowledge and experience between them.

The psychodramatic supervision emphasizes the living relations among the participants, encouraging an opening 'into' themselves and the mutuality between them. The supervisee suffers growing pains in supervision, often leaving him/her vulnerable. This vulnerability stems from the fact that he/she are simultaneously challenged by the roles that express their personal feelings and those that express their roles as students. The supervision offers him/her a combination of a supportive and professional treatment to facilitate both his/her personal and professional growth.

Practicing psychodramatic supervision is a challenging task. The unique social atom, the many and different roles, the double and triple tasks – all these make supervision a complicated process, all the participants in which undergo growth and change.

References

Artzi, E. (2001). *A Supervision Model in Psychodrama*, Doctoral Dissertation. Budapest: Eotvos University.

Hollander, S. L. (1974). Social Atom: An Alternative to Imprisonment. *Group Psychotherapy and Psychodrama*, 27, 173-18.

Holmes, P., Karp M., & Watson, M. (Eds). (1994). *Psychodrama Since Moreno*: Innovation in Theory and Practice. London: Routledge.

Kellerman, P. F. (1992). *Focus on Psychodrama* – The Therapeutic Aspects of Psychodrama, London: Jessica Kingsley Publishers.

Kipper, D. (1967). *Psychotherapy through Clinical Role Playing*. New York: Brunner/Mazel Publishers.

Kron, T. (1994). (In Hebrew) [What is Supervision: Dilemmas, Approaches and Models.] In T. Kron, & H. Yerushalmi (Eds.), *Supervision of Psychotherapy*. Jerusalem: The Magnes Press, The Hebrew University.

Marineau, R., F. (1989). *Jacob Levy Moreno 1889-1974* – Father of Psychodrama and Group Psychotherapy. London and New York: Tavistok/Routledge.

Moreno, J. L. (1961). The Role Concept; a Bridge between Psychiatry and Sociology. *American Journal of Psychiatry, 118*, 518-523.

Moreno, J. L. (1972). *Psychodrama Vol. I.* Beacon, New York: Beacon House.
Moreno, J.L. (1978). *Who Shall Survive?* Foundation of Sociometry, Group Psychotherapy & Psychodrama. Beacon, N.Y.: Beacon House.
Rolef, A., Gerber, M., & Keet, C. (2010). (In Hebrew) [*A shared space: Relational Coaching.*] Internet/Betipolnet – Professional Paper's Section.
Williams, A. (1995). *Visual and Active Supervision, Roles, Focus, Techniques*. New York: Norton & Company.

Trainee´s anxiety to direct: supervision as a journey from anxiety to curiosity

Arşaluys Kayır

In the 1980's when psychodrama training had started in Turkey at the Dr. Abdulkadir Özbek Psychodrama Institute, other training methods of psychotherapy were almost non-existent. Psychodrama had raised attention, but the fact that the training process was done in groups created difficulties. In the advanced group the trainees were expected to direct a play in the training group in the presence of their trainers and supervisors. Anxiety and fear of performance were mentioned informally, but not expressed in the training group. The trainees would avoid leading the group in the presence of the supervisors.

Some trainees never resolved this performance anxiety. Almost half of the first psychodrama trainees of the institute stopped their training. In time, when the effectiveness of psychodrama became more noticeable, those who dropped psychodrama half way through the course expressed words of regret and such phrases like "I could also have done it".

As a trainer and supervisor I observed similar attitudes of avoidance in taking the leading role due to anxiety or shyness among trainees. Therefore I will focus on the role of anxiety in the supervisory process in this chapter. On the basis of my own practical experiences as a trainer and supervisor the issue of anxiety and directing in a training process will be looked at. Furthermore I will propose some efficient techniques which can help to overcome this problem and how to transform dominant feelings of anxiety into self-confidence and curiosity within the supervisee.

1. Anxiety and perfomance in social situations

Anxiety is an emotional state or reaction often characterized by unpleasant feelings of intensity, preoccupation, disturbance, and apprehension (Spielberger, 1972). Most people have experienced performance anxiety in some form and various degrees.

Cognitive symptoms of anxiety include fear of making mistakes and feelings of inadequacy and worrying about things happening. Anxiety in a performance group is inevitable (Landy, 1994). Mastery of task and anxiety are related. High levels of anxiety are related to low levels of confidence (Abel et al., 1990). According to Yerkes and Dodson's (1908) anxiety and performance are related. They could demonstrate that anxiety is an arousal state which – in moderate amounts – serves to motivate the individual and to facilitate task performance. However, if the level of anxiety is too low or too high then motivation and performance is weakend.

In the development of a skill a mild level of anxiety provides positive motivation for developing that skill. It enhances concentration and remembering the experienced moments well. The person learns how to cope with difficulties by using its skills (Nideffer, 1993). But if anxiety increases cognitive performance declines (Eysenck, MacLeod, & Mathews, 1987).

Self-confidence, assertiveness and anxiety are not only characteristics of an individual personality and a specific situation, but also refer to socialisation in a specific culture. In Turkish culture – for instance and generally speaking – assertiveness is not always supported at homes. Eskin compared different cultures regarding this attitude and found a connection to the interpersonal style:"Assertiveness emphasizes an individualistic interpersonal style which is valued in some cultural contexts but not so much in others" Eskin (2003, p.1). He is referring to Kagitçibasi (1996) when he describes the Turkish culture as a 'culture of relatedness', where interdependency is valued. In particular, women are raised to be less assertive. Eskin found out that non-assertive girls are preferred to assertive ones.

This socialisation process probably has an impact on training groups as well. At the aforementioned Özbek Institute trainees are mostly female. Sometimes there are ten women and one man in a training group. Although male trainees could also have difficulties being evaluated by their female supervisors it is more likely for women to feel threatened by that learning situation. However, unassertiveness can lead to difficulties for both genders. To feel anxious or shy can become an obstacle when leading a group under supervision.

2. Performance anxiety and psychotherapy training

Performance anxiety is widely spread in different occupations, where students have to take a leading role in social situations. Social anxiety, performance anxiety, social phobia are concepts to describe this phenomenon. Among professional musicians as well as among music students for example, studies have shown that performance anxiety occurs frequently and can cause considerable distress. However, as already mentioned above, distress can also be a motivator for good performance. Möller (1999) has used "stage fright" for the positively connoted state and "performance anxiety" for disturbing levels of stress.

In psychotherapy distress and emotional experiences very often centre around the dynamics of anxiety. In the infant, anxiety arises whenever the supporting interpersonal relationship fails. Adult anxiety arises whenever the ego breaks down, resulting in a feeling of inadequacy. Anxiety assumes central importance in the whole therapeutic process (Whitaker & Malon, 1981). Self-confidence and assertiveness are necessary to overcome feelings of performance anxiety. Practicing more assertive behaviours in a situation which formerly aroused anxiety and inhibition, can reduce the strength of the anxiety-related response (Wolpe, 1969).

For psychotherapy trainees Bernard (2009) proposes that if there is an optimal level of anxiety for supervisees to experience then supervisors logically have the simultaneous goals of helping to keep their supervisees from engaging in anxiety-avoiding behaviours and of helping to keep supervisee anxiety in bounds so that it works in the service of performance. Approaching to take the leader's role does increase anxiety in supervisees while there is also curiosity in it. At this point the supervisor has to be alert enough to see the hesitations and to offer support. Assertiveness and self-confidence can be considered as those elements that help to reduce anxiety and fear.

3. How to reduce performance anxiety and how to foster self-confidence and assertiveness?

There are different ways of reducing anxiety in a situation where trainees have to perform and are evaluated. Since Moreno (1972) many authors have underlined the importance of developing spontaneity, encouraging creativity and using humour. These components seem to be the most precious keys in the regulation of anxiety and curiosity.

Based on my own observations and feedback which I have collected in our psychodrama training I was looking specifically at the factors, which help to reduce performance anxiety of the supervisee. In a feedback 24 trainees who had completed their psychodrama training and supervision were asked to answer specific question regarding anxiety in becoming psychodrama leaders. Some students mentioned that it is more challenging and it triggers more anxiety to lead in front of one's own training group and to be supervised in the same moment than to bring in a case from the private practice into a supervision group to work on. This feedback was in line with my observations made during the training and supervision. The helpful features for trainees can be clustered under four headings:

- Little steps increase trainee′s self-confidence and assertiveness
- Fostering a supportive relation between supervisor and supervisee
- Providing constructive and sincere feedback
- Personality related aspects within the supervisor

3.1 Little steps increase trainee`s self-confidence and assertiveness

Like psychodrama training, the supervision process itself is done directly with the psychodrama method. When trainees are asked to take on the role as a director in the training group, feelings of anxiety and curiosity are usually present at the same time.

Anxiety produces adrenal stress hormones which, if excessive, reduce the capacity of our hippocampus to integrate new data. This impairs spontaneity. A warm up can help to get back to a more favourable dynamic psycho-physical state. Warming up is essential to reduce anxiety and defences (Hug, 2007). Since psychodrama gives a chance for spontaneity it will accordingly help to regulate the anxiety level and bring it to the curiosity level. Therefore, taking anxiety of the supervisees into consideration it is essential to warm up the group for their specific task, and to get from simple activities and reflections to more demanding ones.

Warm up: As an example of how to start, I would advise the trainees to walk around in the room – as if they were the group leader – and listen to the members of the group. This is a warm up to their new role as a group leader under supervision. Then they are asked who wants to take on the role of a group leader for a next step. A trainee may take on the role and suggests to continue with a warm up play. Whatever is proposed by a trainee can be put into action by the group immediately. Some like to give instructions, while others are more hesitant. Exploring this new role in an easy way takes away the tensions and makes them laugh together.

Director and protagonist in a small group: Then I would take on the role as a director again and I ask them to close their eyes and to focus on what they would like to say today. When eyes are opened again, they are invited to talk to each other. In a next step I would ask them to form groups of three people, taking a role of a director, a protagonist and an antagonist and to enact in little vignettes their play. All groups enact at the same time. The supervisor just walks around and is confirming that whatever is enacted in the group is fine.

In this phase and in the different groups the trainees are visible as directors and actors, but at the same time they are not that much exposed to the whole group. With these small vignettes focusing on their own subjects they warm themselves up to the stage.

Feedback from the small groups: After they have finished, we go back to our big circle and group members give feedbacks. The directors are very curious to hear the protagonist's feelings. Getting used to listen to criticism has to be developed with gradual leading exercises. They ask each other questions. The leaders share their feelings and can hear common difficulties in directing. Some other feedback underlines that it is less anxiety-provoking if all leading is done simultaneously without audience and supervisor watching directly. Changing roles in a small group of three, one after the other motivates them towards directing rather than being protagonist.

Anxiety provoking exercise in the training group: The superviser sits in the circle of the group, and puts an empty chair in the middle of the circle facing herself. By turns each supervisee comes and sits on the chair gazing (watching) the superviser in silence for a minute. After every group has finished each trainee gives feedback about his or her experience during this exercise, waiting for their turn, sitting in the middle, returning back to the circle. With this exercise they realize the differences of increasing and decreasing levels of their anxieties according to the situations. They hear about the similarities and differences among themselves. Feedback suggests that anxiety increases while waiting for their turn, but it decreases when something is performed, and the initial stress is lost.

Group leader in the training group: After these gradual steps we pass into the classical style of live supervision groups done in three levels. The supervisees volunteer for directing, choose their co-leader and lead the training group in pairs. The supervisor–trainer sits outside the circle and does not interfere at all. On this level the supervisees are active.

Feedback from the supervisor: When the session is finished we continue with an evaluation from the leader and supervisor. Then the group members who have taken roles contribute with their feedback. Members are keen to hear from the ones who took roles.

At the end the supervisor gives her/his feedback. Live supervision done here and now will support the supervisees in gaining spontaneity and creativity. We always start by commenting on things that were good and worked well as Fontaine (2001) suggests, beginning with at least three examples which give positive feedback. Then come the weak parts. Feedback about the weak parts is always clearly and frankly given which seems to be very useful, indicating that they can be improved.

3.2 Fostering a supportive relation between supervisor and supervisee

During supervision I continue to be the therapist, educator, group leader, teacher and trainer. My usual relationship with the supervisees is soft and calm. In the words of a supervisee : "When I fall, there is a pillow under the rock". It can also be called strict, but not in a hierarchical meaning. I support and then criticize. I support them in a 'natural' manner, which normalizes the situation. An example of a supervision session (five days workshop, all staying together) can illustrate this way of reducing anxiety in a group:

Supervisor: Have you decided who the next leader is?
Trainee: It should be my turn. Since we have to return tomorrow... but... (The voice goes down and can hardly be heard, hands trembling). I am anxious.
Supervisor: Of what?
Trainee (low voice): Somebody...you... watching from outside.
Supervisor: Do you mean I sit in the circle?
Trainee: No, no, it will be worse. I will, I must...the patients in the group I led liked the session, but here, I'm not sure, but I'll do.
Supervisor sits outside the circle as usual but spontaneously decides to turn her back to the group and starts knitting.
Supervisor: I don't see you but I want to hear you... (Trainee laughs and gets a clearer voice.)
Supervisor: Now I can hear you better.

My supervision style has changed over time. At the beginning I had the tendency to intervene by doubling the supervisees. "When working with a protagonist, the supervisees would put into words their concerns like but what if I cause harm…or miss the most important point? " In a way I was encouraging them to direct and give the feeling "I am here to guide you". For a short time it was helpful. It was not good for keeping their concentration on the work. Now I am visible but give the feeling that I am also invisible, not there.

I respect them because I do not see them only with their 'psychodrama supervisee' identities. I am not only their psychodrama supervisor but I am also interested in their life events. It is a close, natural and motherly relationship. I feel free asking about their life events. I don't judge their life style or their professional work. For our culture this is sometimes a relationship longed for with parents.

3.3 Providing constructive and sincere feedback

In a group supervision, the evaluation and the feedback are not done privately, so I take the responsibility to respect all the members' knowledge and observation. It is better to be frank and share observations.

When a supervisee sits and waits for feedback after directing a session she or he is both anxious about what is coming but at the same time curious [about] what new information she or he will get. In the feedback I put the emphasis not on the directors but on the situations. Criticism is aimed at the behaviour that can be changed. Harsh criticism as well as too much praising is not useful because it does not help the supervisees identify areas for improvement. The consequence is that they are given neither an opportunity to correct their mistakes nor guidance on how they are developing as psychodramatists. Direct, constructive, clear and sincere feedback with appropriate timing provides good guidance and sharing. The supervisees become more motivated to present their development.

The critical point is to give enough information that they can do better because they have certain capacities. For example, I remind them to use their experiences and knowledge they gained wherever they work, in the school, in the clinic, etc. This strengthens their self-confidence that they can use their basic knowledge of other areas, say psychology or psychiatry, in psychodrama where they are less experienced.

It is also useful to make them realize their increasing anxiety signals either on the body, in the mind or on the feelings and teach them alternative ways of approaching tasks. For example, how near or how far they were to the protagonist, what was happening in their posture, mimics. Many short practices are part of the supervision sessions for developing more flexible conditions. However, it is important to keep a focus on how to be flexible, creative, spontaneous, how to use humour and metaphor and be playful, which needs much more time and experience. These features are not as concrete as techniques like role reversal, doubling, mirroring, future projection mirroring, relaxation techniques, soliloquy which are learned more readily and which are useful for balancing anxiety (Figge, 1982).

I do not get angry during supervision, but I give ironic feedback if problematic directing is repeated. On the other hand, when there is something very good, I 'put it into the pockets' of the others. It becomes a 'picnic area of tasting each other's food' and enjoying the time shared together. In that way supervision creates a wider vision for them.

3.4 Personality related aspects within the supervisor

Humour has a warming up effect when it comes surprisingly at moments of stability. The use of humour or relaxing and breathing exercises (imagination) has also proven to be useful, helping to demystify the "fear of having fear" (Stevenson, 1992). Humour provides flexibility to body and mind. It decreases performance anxiety and fear of failure. Humour and imagination are good examples of spontaneity. In a time of crisis one needs laughter.

Good humour is something that comes spontaneously. I like humour in psychodrama because usually we work with pain, and humour helps us to brighten our vision and return to a balanced view of life. Humour enables flexibility and with good timing, it creates a human atmosphere where pains and pleasures are shared together with smiling faces.

Often important metaphors are bound up with humour, which combine the abstract and the concrete in a special way. Humour makes it easier to keep things in one's mind and remember them. Changing the present image in mind also gives time to rest, to laugh, to make connections and associations (Siegelman, 1990). I have the flexibility of finding and using metaphors with quick associations. Sense of humour is important if it has a message in it that clarifies the situation. Metaphors taken from a known song, or proverb or saying, better strengthen the situation to be kept in mind. It is exciting to see that the supervisees learn and borrow things they like from their supervisor. When the supervisor uses and becomes a model, the supervisees also try to use it.

Playfulness and flexibility are also important attitudes for the supervisor as a model. When trainees as 'new' colleagues in the role as a director share their work, then it is for me like playing together. I get in touch with supervisees at a level of affective participation ready to contribute when needed. They can see my tears and hear me laughing. Besides that it is exciting to see how the new directors put theory of psychodrama into practice. This gives an opportunity to see how much I have taught and what more I can teach.

As a supervisor I share whatever I teach by giving hints and information on how I have learned or how it has been a difficult moment for me to bring a good solution out of it. I give examples from other known psychodramatists, how they would approach or had approached such cases. This is a message that conveys that a style for psychodrama is gained in time with certain experiences, opening the eyes and not being blind. I emphasize what is inventive and original, what is new or good for each. This is encouraging and it gives them freedom to create.

Humility: It is good that a supervisor is humble, because the supervisee- supervisor relation is already hierarchical. I also feel relaxed when I transfer what

I know in a natural way. If I feel the pleasure of giving then it is taken in a pleasurable and curious way. The fear of failure or performance anxiety is transformed to "I can do better next time, I will try".

4. A master-apprentice relationship in a journey to curiosity

Psychodrama is full of surprises and peak experiences, therefore it is a 'curious method' on its own. In supervision, I use this tool for better and more enjoyable learning. In psychodrama, there are more unpredictable possibilities than the ones that can be planned beforehand. It is not easy for supervisees to feel self-confident with many unknown aspects in the role of a conductor. A fair amount of anxiety in a group turns the attention into curiosity. As a supervisor I use some moments to throw out some words to wake the supervisees up and make them curious to wonder and want to find out more about what the next step would be. Each person interprets the words according to her/his needs.

During training the trainees are enthusiastic to discover themselves in their psychodrama group. They share their feelings and give good role feedbacks. They are comfortable letting their trainer find out what is going on in the group. However, when their turn comes to direct with the same trainer as the supervisor, there we can see a 'master-apprentice struggle'. The soliloquy of the supervisees is "you can manage better than us" and the supervisor's soliloquy is "go on, you also can do it." The supervisees perceive my alertness, that I follow every detail in silence. After enactment, they are curious to hear about their new roles as psychodrama directors. They appreciate being attentively followed by the supervisor.

5. Conclusion

The function of supervision is supporting, teaching, sharing, guiding, and putting experience in function (Emunah, 1994; Blatner, 2000). When time for directing under supervision comes it is better to meet more often for more practice of directing and give more time for sharings. Anxiety is reduced and learning becomes better when common difficulties are shared and supervisees hear each other's evaluations and criticism one after the other.

The relationship of theory to practice in psychodrama is like a master-apprentice relationship in teaching and learning. There is an interrelation of giving and taking in both. The wish and curiosity of a psychodrama trainee is to become a good leader. A supervisor-trainer is expected to reflect to the supervisee the feeling of love and pleasure to understand human being with a psychodramatic approach.

When writing about my relationship with the supervisees, my style became more concrete and visible for me. It was like a supervision of supervision. Like standing on a chair, I was looking at myself as a supervisor. Do I like it? What do I want to change, to add, to eliminate, to develop? As Fontaine (2001) put it, we are partially blind of our style. Now I am more in touch with my feelings seeing new masters appearing on the psychodrama stage.

References

Abel, J. L., & Larkin, K. T. (1990). Anticipation of performance among musicians: physiological arousal, confidence and state-anxiety. *Psychology of Music*, 18, 171-182.

Bernard, M. J., & Goodyear K. R. (2009). *Fundamentals of Clinical Supervision*. Ohio: Pearson.

Blatner, A. (2000). *Foundations of Psychodrama: History, Theory, and Practice* (4th ed.). New York: Springer.

Emunah, R. (1994). *Acting for real. Drama therapy process, technique, and performance*. New York: Brunner/Mazel.

Eskin, M. (2003). Self-reported assertiveness in Swedish and Turkish adolescents: A cross-cultural comparison. *Scandinavian Journal of Psychology*, 44, 7–12.

Eysenck, M. W., MacLeod, C., & Mathews A. (1987). Cognitive functioning and anxiety. *Psychol. Res.* , 49 (2-3), 189-95.

Figge, P. A. W. (1982). Dramatherapy and social anxiety: results of the use of drama in behavior therapy, *Dramatherapy*, 6, 1, 3-19.

Fontaine, P. (2001). The Development of the trainee's own personal style. In P. Fontaine (Ed), *Psychodrama training. A European view* (pp. 309-316). Louvain (Belgium): FEPTO publications.

Hug, E. (2007). A neuroscience perspective on psychodrama. In C. Baim, J. Burmeister, & M. Maciel. *Psychodrama, Advances in theory and practice*. London: Routledge.

Kagitçibasi, Ç. (1996). *Family and human development across cultures: A view from the other side*. New Jersey: Lawrence Earlbaum.

Landy, R. J. (1999). Role model of dramatherapy supervision. In E. Tselikas-Portmann (Ed.), *Supervision and dramatherapy* (pp. 114-133). London: Jessica Kingsley.

Moreno, J. L. (1972). *Psychodrama*, Volume One, Beacon, NY.: Beacon House.

Möller, H.-J.(1999). Lampenfieber und Aufführungsängste sind nicht dasselbe! (Stage fright and performance anxiety are not the same!) *Üben & Musizieren*, 5, 13-19.

Nideffer, R. M. (1993). Attention Control Training. In R.N. Singer, M. Murphey, & L. K. Tennant (Eds.), *Handbook of research on sport psychology* (pp. 542-556). New York: Macmillan.

Siegelman E.Y. (1990). *Metaphor and meaning in psychotherapy*. New York: The Guilford Press.

Spielberger, C. D. (1972). Anxiety as an emotional state. In C.D. Spielberger (Ed.), *Anxiety, Current Trends In Theory and Research*. New York: Academic Press.

Stevenson, M (1992). Maria: dealing with panic attacks with the help of psychodrama and cognitive therapy. *British Journal of Psychodrama and Sociodrama*, 7, 1, 11-21.

Whitaker, C. A., & Malone, T. P. (1981). *The roots of psychotherapy*. New York: Brunner/ Mazel.

Wolpe, J. (1969). *The practice of behavior therapy*. New York: Pergamon.

Yerkes, R. M., & Dodson, J. D. (1908). The relation of strength of stimulus to rapidity of habit-formation. *Journal of Comparative Neurology and Psychology*, 18, 459-48.

III. Specific Methods in Supervision

Specific methods in supervision – an introduction

Hannes Krall & Jutta Fürst

Generally speaking a method can be described as a repeatable way of doing something in order to reach a goal. Supervision in training aims not only at a specific case which is presented by the supervisee. Supervision should also help to improve the skills and knowledge of the trainee and to reduce the blind spots in general. Specific methods can be seen related to the following aspects:

- setting (individual, pair or group);
- supervision based on observation (live supervision) or on a report (verbal, written, audio- or videotaped);
- structure of time and process;
- supervisory interventions;
- theoretical rationale.

Only for some very easy tasks it can be said that just one method is the pick of the bunch. For example, making fire for a nice barbecue without having matches can be quite challenging. Some experienced people will rely on their practical knowledge. Others have read something about and will put it into practice and the next will simply do it by trial and error.

Supervision can be a way of exploring different options in a co-creative manner. In the case of making fire we would find out that there are some things that has to be respected otherwise we will fail for certain. It is clear that e.g. wet tinder will never give a nice fire. In certain conditions only very experienced fire makers will be able to adapt the method to a particular situation.

If trainees face difficult situations in work and life there are often various ways to deal with it. If we look at the variety of different practices in supervision one can find a lot of different methods and techniques. There are many psychodramatic methods applied in the field of supervision, there are also a lot taken from other traditions, and there are also quite a lot of specific methods and techniques which developed over time in the work of the supervisor or in the respective psychodrama training institution. It can be a helpful stimulus to learn from examples and experiences of professionals who work in different countries under different conditions.

The contributions in the following section focus on different aspects in supervision. Chantal Nève-Hanquet pays significant attention to the repetitive patterns at the workplace and in supervision and shows how to use doubling and role play to help the supervisee to get new insight.

Jan Lap is emphasising the reflection upon the relationship and the dynamics between the director and the protagonist in order to increase the responsibility for the well-being of the protagonist. Based on a five-steps-structure he works on the 'fields of tensions' between conflicting aspects using action techniques and verbal interventions as well.

Ildikó Erdélyi describes examples of her supervisory work and discusses possibilities to combine a protagonist centred psychodramatic approach with a group centred psychoanalytic work based on Ferenc Mèrei in supervision. In two further examples it is explained how the psychodramatic enactment is used in order to reflect the counter transference of a director.

Hilde Gött illustrates her work on stage to strengthen confidence of the trainees in their professional roles and to help them to find their identity as a psychodramatist. In a first examples she focuses on an 'exchange of ideas' among trainees; in a second one she puts emphasis on reflecting different roles of a therapist and how to integrate them.

Psychodrama and role-playing techniques in supervision

Chantal Nève-Hanquet[1]

Group supervision can be an opportunity to experiment with different psychodrama and role-playing techniques in order to reflect upon difficult situations encountered in professional lives. Psychodrama helps to explore multiple perspectives, while allowing elaboration of, and experimentation with different versions of what might or could be possible. The experience of life played out in the 'as if' play enables participants to approach, in all its intensity, their own lived experience and also that of the other (Anzieu, 1975, p. 57).

In this chapter I want to describe two examples to illustrate possibilities of psychodramatic approaches in supervision. The first part will focus on an example of how to deal with problematic repetitive patterns of interactions at the workplace and in supervision. The second part will provide examples of some specific psychodramatic techniques which can be very helpful in supervision practice in working with groups.

1. Repetitive interaction patterns at the workplace – individual supervision

Brigitte trains and supervises a group of about ten telephone-support workers. Jocelyn, a member of this helpline service, told the group that she received a phone call from a man who first asked for help and then immediately insulted her and criticised her incompetence. She felt unable to respond and decided to raise this matter at the next supervision session.

Brigitte, the trainer and supervisor in this group, suggested conducting a role play with Jocelyn as protagonist who is responding to her aggressive client. After this play Sophie, another member of this group, criticised her colleague Jocelyn sharply. Sophie wanted her to tell the client that she does not accept this way of talking to her.

Brigitte – in the role as supervisor – felt that she could not deal with this conflict in a proper way. Therefore she decided to bring this difficult situation up in her own individual supervision, where she first how the conflict between Jocelyn and Sophie had evolved:

---

[1] This chapter has been written in collaboration with Magali Pierre and Brigitte Tilmant

Jocelyn: "I had some difficulty with a call last week. I'd like to work on this with you today." Brigitte checked that the group agreed to the request and proposed a role play: Jocelyn chose Stephane to represent the caller. In this role Stephane shouted and insulted the telephonist: "You're completely useless, you simply don't understand what I'm telling you. In God's name, what's the point in phoning you?" Jocelyn felt very uncomfortable and remained silent. To the rest of the group she said: "I just don't know how to respond".

Brigitte as the supervisor asked the actors to tell how they had experienced this interaction in the role play. In that moment Sophie, one of the group members, burst out addressing Jocelyn in a peremptory manner: "You should have told him that you refused to accept his way of speaking to you!"

Brigitte attempted to calm down the situation by suggesting that the group should first hear Jocelyn's and Stephane's role feedback from this play. Jocelyn: "I felt awful, useless, completely powerless. I didn't dare to say anything in case he just became even angrier." Stephane: "I felt her to be distant, and her silence increased my frustration; I needed her to set me some limits." In this moment Sophie broke in again and addressed Jocelyn sharply a second time: "You should have just said you'd had enough and hung up".

In supervision Brigitte said how ill at ease she had felt on account of the way Sophie had criticised and harangued Jocelyn: "It's not the first time that this participant has reacted like that towards other members of the group. It's not that she doesn't have a point, but her manner is so overbearing and sharp."

The supervisor suggested to Brigitte a replay of the situation experienced in the training session with the use of empty chairs. Each member of the helpline group would be represented by an empty chair and Brigitte would place a post-it on each chair with the names of the participants.

The supervisor then invited Brigitte to double one by one the members of the group in order to understand better their intentions and expectations in that specific situation. When Brigitte was doubling Sophie she was surprised about certain associations she got in that moment. Playing the role of Sophie and her brusque reaction to Jocelyn Brigitte could suddenly find a link to her own biography. She remembered that she had suffered a great deal on account of her father's violent mode of speech. Later on she had to put a lot of effort to free herself from the hold he had exerted over her. Nor had she always had an easy relationship with her sister whom she had considered excessively submissive.

After this reflection in supervision with empty chairs Brigitte stated how much the experience of doubling certain participants had enabled her to come into closer touch with the group members. She was thus able to begin to perceive some of the lateral transfers taking place within this group. She felt strengthened by understanding better the dynamics of the repetition that evolved in her supervision group: Stephane, in the role of the aggressive caller, shouted at Jocelyn, who did not respond. And Sophie shouted at Jocelyn – and implicitly at Brigitte – who reminded her of her passive and silent sister. The supervisor ended the session by asking Brigitte to consider how she might use the results of the supervision session when she will meet the group next time.

2. Complexity of social interactions – role play in group supervision

The following exchange took place during a supervision session with psychodrama practitioners who had attended a training course in role-playing. Stephanie, a psychodrama practitioner and trainer, spoke of her training work with a team of carers employed in a children and young people's home. She brought the following case to supervision:

During a supervision session a social worker whom we will call Lucien, had described his difficulty in relating to one of the residents, Ovid, a boy of 12. Stephanie had proposed the role play of a meeting between Lucien and Ovid. During the enactment Lucien had been unable to communicate with Ovid. Stephanie – in the role of the supervisor – had felt at a loss as to how she might help him.

In her own supervision Stephanie asked for suggestions from other participants as to how she might work on this situation with Lucien in the next session.

With the group we began by considering the different action methods that Stephanie might use. Suggestions included the following: an 'Empathy circle'; a 'Role-playing cascade'; an 'Empathy revealer'.

2.1 The 'Empathy circle'

The 'Empathy circle' is an exercise that can be practised with teams of colleagues which allows participants to gain a feeling of what is experienced by a person called 'P', the subject of the exercise. In the circle of participants, an empty chair is placed next to the group leader. Stephanie could then have asked Lucien to voice his questions/concerns in relation to Ovid. In a second phase, she could have invited him to play Ovid. Lucien would then have come to sit next to Stephanie on the empty chair. The director and the group members would have helped Lucien to feel Ovid's presence by asking questions phrased using the first person pronoun 'I': e.g. How old am I? How long have I been in this home? What kind of a relationship do I have with my parents? Do I have any brothers and sisters? What kind of things do I enjoy sharing with them? What are my activities with Lucien?

It is important that the questions should be open requests for relevant information. The process of asking and responding to questions is designed to open up varying areas of awareness, enabling emergence of an alternative understanding of the dynamic at work in the relationship between Lucien and Ovid.

In this empathy circle Lucien could gain new insights and ideas for his work with Ovid. The fact of having played the role of the other gives rise to empathy in relation to both oneself and the other. Micro-changes are then able to emerge in the relationship.

2.2 The 'Role-playing cascade'

In the Role-playing cascade, a participant (A) is invited to play the role of an absent person (P), with whom he (A) experiences difficulties. Opposite to person (P) are three other participants (C, D, and E) who represent three different attitudes. In this particular case, the point would be for Lucien to encounter differing attitudes displayed by Ovid. The procedure can be as follows:

- Stage 1: P meets C (whom the group leader has asked to be *demanding*); D and E are not present during this interaction.
- Stage 2: P meets D (whom the group leader has asked to be *affectionate*); E is not present at the interaction.
- Stage 3: P meets E (whom the group leader has asked to be *submissive*).

These three interventions are followed by a feedback phase during which C, D, and E describe their experiences in their roles. A, in the role of P, first describes what he experienced with the three different characters. A then reflects about the insights which he gained from that interaction. He also reflects to which aspect he will pay particular attention to during a future meeting with P.

2.3 The 'Empathy revealer'

In the supervision Stephanie finally chose the 'Empathy revealer' as the action method most appropriate for her situation. With this approach the perspectives of the teams of social workers on the interaction between Lucien and the 12 year old Ovid are to be explored.

Stage 1: Choosing roles

Stephanie wrote on pieces of paper the names and positions of each of the social workers, which she is supervising. Each colleague in her supervision group is choosing one name of a social worker. André, a colleague of Stephanie in the supervision group, got the role of Lucien who is working with Ovid.

Stage 2: Double role changes

Two chairs were placed on the psychodrama stage, one representing a social worker, the other one representing Ovid. The perspective of each social worker – represented by different colleagues – regarding the relationship between Lucien and Ovid will be explored by the following steps.

The supervisor asks all participants in this play to perform a double role change: Firstly they change into the role of a social worker, secondly they are asked to revers roles with the boy Ovid. In this way they can develop empathy in both roles.

André, who choose the role of Lucien, first gained empathy with the role of the social worker, and later on he reversed roles with Ovid. By doing that André at first develops empathy with Lucien, and then by having this insight from Lucien´s role he reversed roles with Ovid in order to get the other perspective as well.

In the role of Ovid he talks about his experiences with Lucien. The supervisor and director helped André to express himself in the role of Ovid, with questions like: "On what sort of occasions do I, Ovid, meet Lucien? What do I like about Lucien? What are the things that bother me about Lucien? In what types of circumstance has Lucien helped me?"

After each sequence of the role play, the supervisor turned around the empty chair for the social worker to signify that another member of the group was about to play Ovid.

Stage 3: Rolefeedback and sharing

After all the roles had been played, there was time for a feedback in three steps: First of all, each participant spoke of his experience in the role of Ovid. The supervisor asked the participants: "When you were playing Ovid, what did you feel?" André – for example being aware of the insight from Lucien's role – said: "I felt how prone Ovid is to loneliness".

Secondly, each participant described from the perspective as a social worker to which aspect regarding the need of Ovid they would put particular attention to. The supervisor asked all who played roles as social workers: "When you see Ovid again, what you will pay particular attention to?" André, playing Lucien, said: "I am aware of the fact that I do not give enough space to Ovid and that I could pay more attention to the time I spend with him at bedtime."

Thirdly, each participant, representing himself, spoke about the points that struck him or her personally during the role play. The supervisor asked: "While you were playing the role of Ovid, were you in touch with your own experience?" André – for example – expressed surprise at discovering a personal memory in this way. He remembered that when he had been sent as a child to a holiday camp he was lonely at bedtime.

Stage 4: Feedback from the supervisee

All these feedbacks and the experiences of the participants in this role play were important to Stephanie, who is supervising this group of social workers. It was especially important what she learnt about the relationship between Lucien and Ovid: "I realise how isolated Lucien can feel. In the role play I was better able to feel the kind of demands that Ovid makes. In my role as supervisor I will do some work on isolation with the group of social workers".

In every interaction, numerous and wide-ranging factors are at play. The enactment can promote the unfolding of these factors and of representations that will be named in the group's feedback. This procedure thus serves to *reveal* the complexity inherent in the reality of situations, lived experience and interpretative grids.

The notion of 'empathy' was experienced in different situations and in different roles: the role of the care recipient (here, Ovid) and of the social workers (here, Lucien and his colleagues). The group participants found that memories of personal experiences come up within them and that this experience helps them to connect with attitudes that they have encountered in their professional practice.

"Empathy is simultaneously a pre-reflective impulse and a skill that facilitates the processing of what emerges in the immediacy of the relationship with the other" (Vannotti & Berrini, 2010, p.109). These authors also state that to approach one another with empathy means a deliberate enactment of one's understanding of that other and of his experience, asking oneself questions about how the other manifests emotions, and seeking to work on the emotion by which we have been gripped. Empathy then fosters the construction of reliability within the relationship.

3. Conclusions

People come to supervision with their spontaneous first-hand perceptions of problematic situations. They experiment by playing the roles of persons other than themselves who are involved in the same situation. This method enables the emergence and identification of different forms of resonance (Elkaïm, 1996). Participants are invited to pay attention to new elements within their own being; these will be linked to their own history, while also having meaning and purpose within the supervision group. The reverberation of this personal experience is thus a level at which work can be carried out and linked with the experience of other group members. In this way, persons in supervision can effect change within themselves at the same time as they work to help other people and/or systems to change.

Similarities and implicit rules between what happened in the representation and the institutional situation presented by the supervisors can be pointed out. Thus, an important advantage of this admittedly rather complex mechanism is the 'gap' that it serves to create. As explained by Christine Vander Borght and Chantal Nève Hanquet: "The gap created by the sequence-within-a-sequence time/space for representation allows participants, above all, to experience full immersion in the emotions and sensations associated with being present in a given place and participating in the interactions taking place there" (1999, p.111).

When people express themselves on their own behalf after the role-playing, they retain the newly gained insights and a process of change can be initiated. Seeking to alter the constructions of the world, each participant enters into a dynamic that is non-judgemental. It is thus that micro-changes in relationships become perceptible in the wake of such experimentation: Changing roles enables the person to appropriate alternative representations, facilitating the creation of a relationship that will be able to move and develop in the light of new insights gained.

References

Anzieu, D. (1975). *Le groupe et l'inconscient.* Paris: Dunod.
Elkaïm, M. (1996). Pour une lecture circulaire de la résonance. *Génération*, 7.
Vander Borght, Ch. & Neve-Hanquet, Ch. (1999). Mettre le corps en jeu. *Cahiers critiques de thérapie familiale et de pratiques de réseaux. 22* Brussels: De Boeck, 103-112.
Vannotti, M., & Berrini, R. (2010). L'essence relationnelle. Fonctionnement neurobiologique et relations humaines. *Cahiers critiques de thérapie familiale et de pratiques de réseaux, 43*, 97-116.

Supervising the interaction between director and protagonist

Jan Lap

This chapter will focus on how supervision is practiced at LapStreur Psychodrama Education Institute. The main objective is to reflect upon the relationship and the dynamics between the director and the protagonist in order to increase the responsibility for the well-being of the protagonist, which is partly passed on to the director, and finally gets shared by the supervisor.

The director has to deal with all the strains a protagonist experiences. Therefore different 'fields of tensions' in the psychodrama practice need to be addressed. The supervisor reflects with the director-supervisee on the choices made in his or her psychodrama practice.

This chapter first of all outlines the basic principles in supervision at our institute and secondly it will examine the options the director has in dealing with specific challenges and tensions.

1. Introduction – basic assumptions and procedures in supervision

Supervision is an essential part of our psychodrama training. It is important to us to do supervision in an active psychodramatic form as often as possible. The objective is to work on the problem as psychodrama directors and not on the problem or the protagonist. However, sometimes the problems of both influence each other. According to the need and the level of the trainee we distinguish between different forms of supervision:

- Live supervision or 'direct supervision': according to the need of the trainee the supervisor can just observe the directing or if necessary intervene. In both cases the supervisee gets feedback immediately after each session.
- Supervision on the basis of a videotaped psychodrama session: the sessions will be looked at on the tape and analysed by using active forms of re-enactment.
- Supervision based on a verbal or written verbatim report (previously sent): parts of the psychodrama are selected for reflection by the supervisor and/or the supervisee and enacted. Different psychodramatic methods and techniques can be used to review the actions and the attitudes of the director.

A psychodrama supervisor acts in different roles. Simultaneously the supervisor will take the role of the facilitator, the consultant (or trainer), the teacher and the evaluator (Williams, 1995). It depends on how much room he allows to his supervisee-director, consistent with the context and the task of the supervision. Beginning psychodramatists probably want a more close-by supervision. In that case the role of a teacher, who guides the director, will be dominant. Later on the supervisor takes the role of a consultant or partner in considering or taking the right actions.

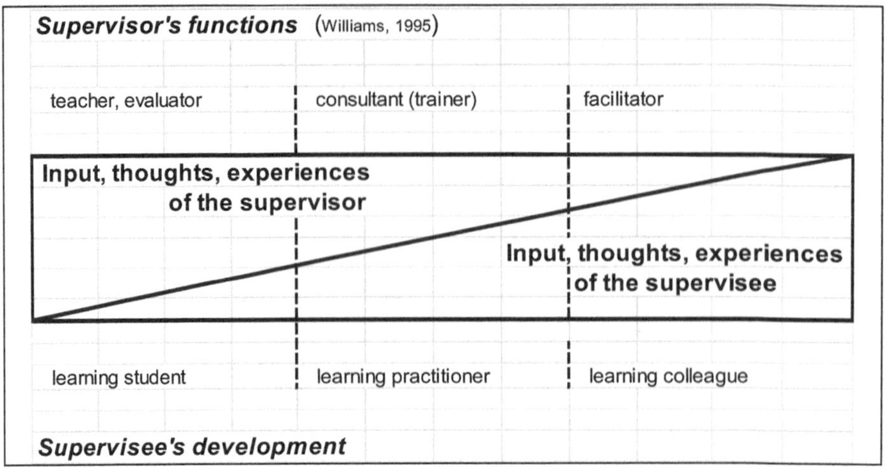

Figure 1: scheme developed and used at LapStreur Psychodrama Education

Our supervision education and practice is based on a structure, known as the Dutch supervision tradition (Kessel, 1993, p. 5-27). It contains the person, (vision on) the profession and the professional's work context, which are connected with thinking, feeling and acting. All of those we bring on stage in psychodrama-supervision (Lap, 2005, p. 1-3). Wherever we can, we follow the next five steps to reach a higher direction performance on stage. This is an application for psychodrama of Kolb's Learning Cycle (Smith, 2001):

Step 1: *Re-enact: Recalling the facts.* Here we are looking for facts (place, time, persons, and report of acts) in action; also the feelings there and then are considered to be facts.

Step 2: *Reflect: Realizing the impact.* On this step we are focusing on the feelings, emotions, thoughts and estimations in the here and now. It is a research on the relived experience. We call this: reflection.

Step 3: *Connect: Gaining insights*. We encourage the supervisees to draw lines to known theories, to their own biography, their present life events and relationships, their hang ups, their projections, (counter-) transferences, etc., and to connect with others, themself, their roots, favourites and their knowledge.

Step 4: *Prospect: Anchoring.* Here is the goal to draw conclusions and plan for the future. The supervisees should take at least one very concrete intention or goal along to their upcoming psychodrama sessions and report about it in the next supervision. Do a try out, here in the supervision context.

Step 5*: Collect: Processing.* In order to help the trainees to process the supervision session we ask them for a one-page written paper about it.

2. The protagonist in a 'field of tensions'

It can be helpful to a supervisor to look at different factors that a protagonist is struggling with. These factors can be seen as a 'field of tensions'. The supervisor should consider, that the protagonist is somewhere on the tension line in between of conflicting aspects. From our long lasting practice we created the following dichotomous structure: bravery versus fearfulness of the protagonist, the talker versus the mover, controlling versus confiding, hanging onto reality versus transcending reality, inter-psychic versus intra-psychic work.

Bravery behaviour versus fearfulness: For the protagonist it takes courage to start working in psychodrama. Whatever the theme is about, a protagonist gives others the opportunity to take a look in his intimate space. Many people have difficulties to expose their problems. Therefore almost every time a protagonist starts his psychodrama, he will experience a certain feeling of hesitation. Paradoxically some protagonists seem to express it by overacting: "We will fix this easily". Other protagonists start cautious and shy. This behaviour can be seen as an expression of feeling uncomfortable, standing in front of the group and not being well prepared.

The rational talker versus the emotional mover: There are protagonists who talk effortlessly. They do it reasoning, seeking, stating, questioning, postponing decisions and sometimes totally self-directing. The psychodrama director seems to become unemployed. One inclines to think this protagonist does this overwhelming talking on purpose. It is really not the case he wants to be in control but it belongs to the life style of such a protagonist either to cover up deeper layers with a lot of words or getting grip on a problem just by understanding and comprehension, assembled in words.

A so called protagonist-mover goes after action (excitement, getting moved, commotion, emotion, lot of facial expressions, laughter, tears). In every action is turmoil, often structureless. He feels like a fish in water if the director fast jumps to action, interaction and playing. It seems easy for a director. Yet the life style of a protagonist of avoiding grip, planning and structure often is a hindrance to move on in his own real life.

Controlling versus confiding: a protagonist is moving permanently on the line between relying on the director and directing his own drama. This movement can get stuck in each one of the poles. Self-directing includes the risk that the protagonist does not pass the boarder he reached already before the psychodrama; he is only re-experiencing what he did before. When something new emerges, it mostly comes from the cooperation of the director or group members. Other protagonists simply rely on the management of the director. They themselves hardly come with any input; they just accept what is offered, directly from the director or via doubling of the group members. It can result in moving psychodramas. However, questions arise: Whose drama is played? To what degree is his creativity mobilised? What has the protagonist learned?

Hanging onto reality versus surplus reality: By surplus reality is meant, that the protagonist has the guts to experiment with unknown impulses, which are handed over from outside. He dares to confront himself with areas that he has avoided or denied (for instance: his so called shadow side). It may suffice sometimes just to face reality and to explore its ins and outs. It is not simply for neither the protagonist nor the director to distinguish which one of both (or perhaps the two) approaches is adequate and productive in the ongoing session.

Interpsychic versus intrapsychic work: The protagonist sometimes wants to find out what exactly happens between himself and a relevant other person (e.g. partner, boss, employee, physician, neighbour). He questions him or herself: "What kind of dynamics influences their interaction?" or "Which behaviour could change the atmosphere?" Further than such questions the protagonist does not want to go (yet).

Others on the contrary want to explore where the behaviours come from, what influences from the past or which inner forces in the present play a role or how to get along with my inner self? Often both sides of this line are entangled, and one aspect can illuminate the other one. It may be important to set limits: either an interaction-drama (Cuvelier, 1976) introduced this term for dramas where social behaviour and relationships are explored) or an intrapsychic psychodrama. Lots of reasons may determine such a choice.

3. The psychodrama director in a 'field of tensions'

The director, and all the more a new director, experiences with every new protagonist a mix of feelings. There may be: curiosity, expectation, and confidence in a well developing psychodrama. Sometimes from the very beginning a protagonist evokes in the director feelings of being wary, fear and defence. Besides that there is always quite a bit self-confidence and professional pride.

A director makes a lot of explicit or implicit choices during the psychodrama. Therefore standing still at the made choices from time to time is an adequate action. Getting detached from grinded processes, preventing repetitive behaviour, freeing from patterns and a push to permanent learning. Finally such a reflection is favourable for his present and future protagonists. The five tension lines are our starting points.

3.1 Managing bravery and fearfulness

The director knows every protagonist has to overcome something in order to step into the spotlights. He will be extra alert, when a protagonist exposes how much he is at ease. From here a performance starts in which the protagonist is going to lose more and more of his self-confidence. The veiled insecurity looses gradually its curtain, which can be humiliating for the protagonist. By bravery hidden insecurity blocks the spontaneity. It is just in the insecurity that one finds the well to explore something fresh and new. The director can help such a protagonist to get closer to his insecurity and sources. A few options:

- The director may show his own insecurity with sentences like: "We enter together this adventure and we don't know where we are going to come - neither I nor you. We'll see…"
- The director names the behaviour of the protagonist and provides a space to change it by telling: "You really like to know what is supposed to go happen; sometimes everything comes about differently; take your chances…"
- The director ignores the protagonist's behaviour and stimulates from the very beginning the group members to double with e.g. following sentence: "I invite you, group members, to double as much as you can. For our protagonist it will broaden his perspective…"

A totally other kind of protagonist shows his insecurity in an evident manner. In that way the protagonist makes an appeal to the director's competence to set the protagonist at ease. The director will not disqualify this insecurity, since he knows there is the space to fill with creativity. He will provide a structure in which the person's issues can be investigated. In order for the insecurity to turn to good account the director has several possibilities. A few options: The director wants to affirm the protagonist by telling him: "It takes courage to set this step, and you have already set a big step by entering the stage!" He can say: "Don't worry about the process; for that matter you have a director!" Also in this case the support of early doubling by the group members can be very encouraging. The supervisor can pick out one of the choices of his supervisee, help him to reflect on it and probably to consider alternatives.

3.2 Managing the talker and the mover

Resentment about protagonists, who just don't stop talking, costs a lot of power from a director. The talking mostly is unexciting, without emotions and absorbing energy. The pitfall is, that a director goes with this verbiage. That counts the more, when a director himself has a preference to deal with problems by reasoning and talking. When the protagonist gets offered what psychodrama really is about, action and performing, he will encounter a lot of new experiences.

The protagonist-talker usually dealt already with counsellors, coaches and maybe psychotherapists. Possibly with success too: his problems are arranged, he gained insight in his issues and received evidence for an approach.

Psychodrama should offer him an extra value. The director can help him. Here a few attempts:

- The director reminds him that action is the way of working here, and immediately he proceeds finding a scene to work on: "You know we love action. That is what we are going to do with you. Can you think of any situation, that shows how your issue is working?"
- The director seduces the protagonist in another direction. He stops the interview and says: "Select a group member who can stand for the one you have something to work out with. Say to him what you just told us about him during the interview."
- The director asks him to sculpt what bothers him, and how severely: "O.k., we know enough. Use everything in this room to show what keeps you busy, and how much it oppresses you."

If a director has to work with a protagonist who slips fast into emotions – laughter, tears, anger, disappointment – action will follow fast: spontaneous vehemence, participation of group members and every one in motion. The director in turn has to be aware, that the protagonist possibly does, what he used to do always and what he already can very well: acting out. Several protagonist-movers, acting out protagonists, come to psychodrama to get channelled to a world with more structure and logic than he lives in until now. His emotions are running away with him, his environment is chaotic and his inner life also; everybody can read his moods on his face, never any rest in his internal voices. Although such a protagonist-mover initially seems easy for a director, he must count with all these just called possibilities.

A golden psychodrama is the psychodrama in which a protagonist makes a connection between feeling and mind. Words usually are the connectors between feeling and thinking. It is evident the director has to do the right interventions in these situations. Here one finds a couple of examples:

- At lively and restless behaviour of the protagonist the director can ask: "I see you walking up and down, fast and pawing the ground. Tell us: what words are coming up and say them loud out."
- With a protagonist, who is expressing his feelings non-verbally (for instance bending over, his hands on his head, crying) the director says: "Stay with your feelings right now, feel them…And then put into words how you feel and what you feel." Also doubling will help in these cases.
- Sometimes an action ends up with emotional statements, like: "Tomorrow I'll go after him and tell him the truth; he will feel himself where I go through!" The director has several options. He can say: "Let us have a look how this will take place tomorrow". A kind of behaviour training. Another approach is focussing on his inner order and suggesting: "Let us line up all the factors that play a role in this statement. We go trying to find a hierarchy in those factors." One can use here chairs, auxiliaries, distance and closeness, musical instruments, etc.

The supervisor can confront the director-supervisee with his choices, and help him to examine whose preferences he or she followed, his/her own or those of the protagonist and why.

3.3 Managing controlling and confiding

An extreme form of self-directing by the protagonist is the autodrama, while a director is present, but does not intervene. The protagonist does everything himself. The director should have a good reason for allowing a protagonist to use this form. Both the protagonist and the director have agreed on this procedure.

A different situation occurs when a protagonist in many ways rejects all director's proposals. That doesn't necessarily happen in a nasty way. It just happens. The protagonist tries to be in control over whatever takes place.

The director, aware of this dynamics, can choose to enter the 'match' or can

ignore it. Each intervention should have the intention to stimulate the self-confidence of the protagonist and his trust in the director, so that he is able to commit himself to the psychodrama process. The director is the preserver and the guard of that process. A few examples of interventions by the director:
• Meta-communication: "Wait a moment, protagonist, what is basically happening between the two of us? What do you expect from me? What happens to you right now?"
• Formally: "Listen, I am the director here. The content is your business, but I am the boss when design and form are involved. I am willing to share with you how we are going to set up the scene."
• Paradoxically: "The group members and me, we stay outside of the stage. You know best what must be done here."

A certain degree of self-steering, fitting the protagonist is desirable. Each director knows and fears the protagonist, who hangs on to him, rushes in his arms: no own input at all, agreement with every proposal, accepting each double, never an own initiative. All the group members including the director and co-director work hard, except the protagonist. To mobilise a protagonist, who makes himself dependent, is a hard job. It needs inventive interventions, which fit to the individual person. Some examples:
• The director places the doubles in front of the protagonist and invites the group members to pronounce their doubles combined with bodily expressions; the protagonist must firstly repeat the accepted double and the expressions also. Then he is invited to say it in his own words (getting acquainted).
• The director places and keeps the protagonist frequently in the antagonist role.
• The director helps the protagonist to explore, whether he behaves in normal life in the same way, and if he really wants that: "Do you recognize 'this always saying Yes' in your daily life? Do you want to keep it that way?" And "Show us a few initiatives you recently have undertaken" (self-esteem).

The supervisor helps the director-supervisee to reflect on his own compulsion and generosity towards the protagonist.

3.4 Managing hanging onto reality and surplus reality

Many protagonists use psychodrama to get better control over some ins and outs in their actual reality. They look for insights and possibilities to handle them. They explore what they have done and the results of it. They identify well known inner voices, and give them a new assignment. They investigate past and existing relationships and search for their significance. The director leads all of this, and has anyhow decided that it has to be that way.

Sometimes he has his reason to decide something different. Reasons may be that the protagonist is repeating often a particular behaviour or is rejecting any double that gives a new impulse or the drama and probably the whole life of this protagonist is very boring. Here are a few interventions, which help a protagonist to break through the wall of his reality:

- Ex machina interventions: "Certainly you have a very sparkling uncle or niece or friend. I propose to put him or her over there. He watched your actions here. What would he say to you about it? What is his advice? Become that person and speak aloud what that person says to you". A variant would be: The Creator, the innovator, God, has seen here everything. On this table is his observation post. Reverse roles with God. God, how do you comment on this life style?" (Deus ex machina)
- Speculum magicum interventions: "We give you the opportunity to look in a magic mirror. Sit down right here. The group members make one minute each free improvisation on parts of your action until now. When you yourself incline to join them, you may improvise on your own actions, like they do."
- Imaginatio-interventions: to these interventions belong: forced fit (two realities are forcefully connected, for instance: you have your business advised by a chicken; a pilot helps you at mountaineering); playing your own shadow-side; magic box; magic shop; best-possible scenario; magnifying; an eulogy for you in superlatives.

Other protagonists prefer to experiment. They try to find alternatives and new versions for all the stable lines in their life. When they have a good relationship, they want to find out what it means to be unfaithful; they enjoy a healthy sleep and then want to discover how it is to suffer from insomnia; they have a conciliatory nature and want to experiment with conflict and hostile behaviour.

A director assesses the reality values of these experiments. He can conclude this behaviour has no extra value for the protagonist. It rather appears to be an avoidance of engagement or confrontation with real life. When one works with surplus reality, it has to serve an aimed change or a future goal. The director will strongly focus on that. A few examples:

- A protagonist wants to observe his own funeral. The director: "I am willing to work this out with you, but I have a condition. We are going to look for forms to shape your funeral and we involve your survivors, like family, kids, friends. We do two things: we help them with choices around your farewell ceremony and we help you to use the possibility to personalise your funeral with an important input from yourself. I don't support immersing in self-pity and self-glorifying!"

• A protagonist wants to investigate how he will react in the case he gets cancer. The director: "Is there in your family or your circle of acquaintances someone who suffers from cancer? Let's first of all interview this person!"
• A protagonist says: "I really want trying to fly. I don't mind to hurt myself, although I am not crazy." The director reacts paradoxically: "O.K., choose someone of the group, who can play you, and is willing to get thrown from a chair we put on top of this table…"

It is a challenge for the director-supervisee to think about the influence he exerted on the protagonist in his dealing with or transcending of the reality. And how he did it! The supervisor will question him about that.

3.5 Managing inter-psychic and intra-psychic work

We are going to work with what happens between the protagonist and his relevant others. On the other hand in the case of intra-psychic work we are going to explore what takes place within the protagonist. We look for options, inner motions, streams, voices, etc. What we choose is obviously not asked for in advance. The interview process and the formulation of the research question (if any) will lead to the first enactment. The drama develops from there and can go in one or the other direction. Sometimes both forms are used.

Reviewing a psychodrama many directors realise the action ended up in one of the forms. Sometimes a director has an unreasoned preference just for one form. Reflection with his supervisor can bring him to that insight. Since both forms are equally appropriate for deepening, it could be that a protagonist profits better from the form that just isn't the director's preferred one.

An interaction drama with the contribution of doubles can touch the emotional layers, sharpen goals, clear relationships, and in its own way direct to personal depths. Similarly an intrapsychic drama, with sub-personalities, inner voices and rearrangement of internal entities can lead to renewal and deepening experiences. Often one form works as a warm up to the other. When do we explicitly choose an interaction drama? A couple of possibilities:
• Training of social skills: the director knows this protagonist benefits the most at this moment of a role-training-like approach, because the lack of social skills evidently hinders him.
• Diagnostic use: The protagonist is confronted with his own empathic ability. The director becomes aware that the protagonist isn't able to reverse roles (clients with particular syndromes like schizoid, psychopathic, with impediments in the autism spectrum have troubles with empathising in others).

• Social atom research: The protagonist first of all wants to know how his closest environment influences his life.

Indications for the use of an intra-psychic drama have to do with indistinct mental positioning, inner doubts and dilemmas, or traumatic early or recent experiences of the protagonist. The contexts can be very diverse. Some examples:
• The protagonist is not able to bring any hierarchy in his preferences in life (e.g.: in choosing further education, type of spending holidays or having kids).
• The protagonist must take a decision with some ethical connotation (e.g.: moving to a nicer place or staying near to his grandchildren; respecting his 18 year old son's autonomy or controlling him; helping at night in a minor car accident or avoiding the risk to be robbed).
• The protagonist wishes to discover why the same pain bothers him over and over (e.g.: movies, where children get harmed; being reminded of first spouse).

The supervisor stimulates counting all the profits of the chosen approach. They may look for alternatives. And both explore the question: what value or benefit another possibility would have generated.

4. Summary

How to learn to manage the described "tensions lines" is one of the main tasks for the director and the protagonist. The supervisor has to manage the tension lines in his or her work as well. Therefore – at least from my point of view – supervision in psychodrama turns out to be a "special job". It is a...
• *responsible job:* In a remote way supervision of a psychodrama-director is sharing the responsibility for his protagonists, mostly his future protagonists. Of course, the director-supervisee is the supervisor's direct and primarily protagonist. In a sense however the psychodrama supervisor carries a double responsibility. Most supervisors are aware of that position. Some of them count only with the direct relationship: "I cannot be blamed for an omission or even harm, caused by a director, when I wasn't there!" It is not that incidental omission or harm. It is not about legal accountability. This responsibility is a moral one: like we are proud of our directors, when we hear from excellent performances and creative dramas, and attribute those results to our education and supervision, so we have to realise that bad... the reader may finish the sentence!
• *dramatic job:* Regrettably often psychodrama supervision takes place without psychodramatic action (Williams, 1995). Frequently the method of counselling is used, sometimes with confronting interventions. Psychodrama supervision is

an elected opportunity to do psychodrama: the supervisor makes use of as many re-enactment chances as possible, particularly in the search for alternatives. It is an art to make the supervisee reflect on a couple of key actions in the psychodrama: he puts them back on stage, evaluates his choices and the reasons behind them, recalls the impact on the protagonist, attempts alternative actions, under direction of the supervisor.

• *nurturing job:* The purpose of the reflection on the executed actions and the alternatives is not to enlarge the action skills of the supervisee. Likely it is a nice side effect. The authentic goal of supervision is to make the supervisee grow as a person within and via his profession as psychodrama-director. The effect is, he increasingly approaches his protagonists in an authentic way, and is able, more and more, to distance himself from his own hang ups and preferences.

• *thrilling job: The* supervisor directs within the psychodrama-supervision re-enactments and alternative approaches; he is then psychodrama-director himself. He has in that role an exemplary function for his supervisees. That makes psychodrama-supervision exciting and thrilling!

References

Blatner, A. (1996). *Acting-in: Practical application of psychodramatic methods* (3rd ed.). New York: Springer.

Cuvelier, F. (1976). Psychodrama en interactiedrama, sociodrama en roltraining. *Tijdschrift voor Psychotherapie*, 2 (1), 207-214.

Kellermann, P. F. (1992). *Focus on psychodrama: The therapeutic aspects of psychodrama.* London: Jessica Kingsley.

Kessel, L. van & Haan, D. (1993a). The Dutch concept of supervision. Its essential characteristics as a conceptual framework. *The Clinical Supervisor*, 11(1) 5-27.

Kessel, L.van & Haan, D. (1993b). The intended way of learning in supervision seen as a model. *The Clinical Supervisor*, 11(1) 29-44.

Kolb, D. A. (1984). *Experiential Learning*, Englewood Cliffs, N.J.: Prentice Hall.

Laat, P. de (2008). Quotations. In Lap-Streur, M. (Ed.), *Conference and Souvenir book*, Heeze, pp. 30-36.

Lap, J. (2005). *Psychodrama und Klinische Seelsorgeausbildung*. Einführung, Kongress Supervision und KSA. Strassburg, p. 1-3.

Moreno, J. L. (1977). *Psychodrama volume I*, (4th ed.). Beacon, N.Y., p. a-e.

Smith, M. K. (2001). David A. Kolb on experiential learning. *The encyclopaedia of informal education.* Retrieved [sept 4, 2010] from http://www.infed.org/b-explrn.htm

Williams, A. (1995). *Visual and active Supervision, Roles, Focus, Techniques*. New York/London: W.W. Norton & Company.

Psychodrama and psychodynamic contributions to supervision

Ildikó Erdélyi

In this chapter I would like to give an account of some lessons learnt during my two-decade-long work in Hungary as a psychodrama supervisor. Among these I will discuss the use of both the protagonist-centred and the group-centred procedures applied in supervision.

By protagonist-centred psychodrama I mean the working method introduced by Moreno (1965). I call the procedure elaborated by Mérei and his colleagues (1987; Erdélyi, 2010) group-centred psychodrama.

In my study I examine how the combination of these two methods can enhance the effectiveness of the supervision and how the psychodynamic approach and analytical method taken from dynamic psychology, or even from psychoanalysis, contributes to it.

1. Objectives and procedures in supervision

I conduct supervision in the following forms: (1) Life supervision in advanced training groups that are held once a month in 10-hour blocks. One third of each block is devoted to group supervision. (2) Individual supervision for advanced trainees and supervision for pairs of group leaders. (3) Reflecting on cases of trainees in the form of a workshop. The themes of the workshop are optional. The trainees can choose a) group dynamics, b) a crisis theme, c) a clinical theme. The themes of the workshops can be combinations as well, e.g. crisis and group dynamics.

I lead workshops in all three themes, combining them if possible. We can process the clinical, crisis and even the group-dynamical theme effectively using the participants' cases. If it is announced in the course description that the theme is going to be processed within the frames of supervision, supervisor candidates are also keen to take part. I took the following supervision cases, in which I describe and analyse my supervisory work from my own records.

2. Psychodynamic approaches in psychodrama

2.1. 'The house': Supervision on a psychodrama group held at a hospital's psychiatry

In the following example two female group leaders are conducting a so-called open psychodrama group at a hospital's psychiatric unit. The two psychologists use tales in the warm-up part of the group sessions. They make their patients connect with their intra-psychic problems through tales.

The group leaders' question in supervision is: "Is the group-centred drama technique an appropriate treatment method in the psychotherapeutic work done in a hospital`s psychiatry?" The group leaders were afraid, that they cannot work safely with protagonist-centred psychodrama in the group where there are patients with splitting defence mechanism (Kernberg, 1979; Holmes, 1992). We experienced that the play's 'as if' space and the group's space could easily melt together for borderline patients. They do not separate fantasy from reality either. The group-centred method tunes the participants on the psychological work with the help of tales, and creates a bridge between the tale and their own problem by means of analogy.

On the supervisor's suggestion the group leaders present their method with telling a tale and putting the theme formed by the tale into a play. In supervision a specific focus in this group is on Isabelle, a patient with an obsessive-compulsive disorder and a borderline pathology.

The tale is created like this: "Once upon a time, there was…", one of the psychodrama directors begins the tale. Then a male patient continues immediately: "...the end of the world", then some female patients go on: "...and there was a house", "...and there was a well in the yard". Isabelle brings the female main character into the story: "...and there lived a girl", then a man again: "...and then a wanderer came", a young woman adds: "...and the storm began", and the man who was the first to speak says: "...and the man stood under the roof".

The patients take up the roles spontaneously. The man who was the first in the story-telling says that he will be the end of the world and he goes 'on stage'. Isabelle takes the girl's role, two women act the house holding their arms up and touching each other`s palms. The man who joined the story-telling as second becomes the wanderer, and two men take on the well's and the roof's roles as well.

The drama halts at this point in the group. Choosing the roles went smoothly, but the work gets stuck in the phase of playing the fantasy roles. In the supervision we continue the play based on the fantasies of the two directors to see what psychodynamic processes might appear in those patients who took on the roles.

'The desire and prohibition of seduction' – a psychodrama in supervision: In the supervision the psychodrama group session is enacted. The 'actors' are on stage. The wanderer knocks on Isabelle's door. Isabelle's role is being played by the two directors in turns. The girl gives the man food, but does not let him in the house, only in the yard, and sends him to the well. Then the storm comes which is being represented by a female participant of the supervision group spontaneously. The wanderer is shivering at the well, but the girl does not regard him. At this point a male participant jumps on stage and says: 'I am the roof', and the wanderer stands below it.

In the meantime other participants in the supervision group create a concept and they put it in play: They build another house at the 'world's end' with a woman living in it. This woman calls the wanderer in and she manages to invite the girl from the other house as well, who remains shy but accepts the invitation.

The group leaders can feel Isabelle´s inner conflict by taking on her role in the play. Their interpretation is, that Isabelle hides her desire-fantasy from her girlhood which she still fears to become real: She could seduce her mother's boyfriend, who replaces her father. Her desire for seduction is enough for her to feel guilty. Isabelle protects herself from that fantasy – which is almost the deed itself for her – with sentencing herself to be motionless.

The reflection on the transference and countertransference of the group leaders in the play suggests, that the leader of the patient group becomes the lenient, 'good' mother (Klein 1955/1986; Winnicott, 1971), and the co-leader is the 'bad' mother, who turns away from her daughter, making her feel guilty. In the countertransference the co-leader feels her own angry mother coming alive in herself during the play. Splitting manifests itself also between the leaders as an effect of the borderline pathology.

In the next supervision session of the advanced group the leaders of the patients` group tell that they managed to direct the drama about the house. Isabelle entered the other house and tried getting free. Her action can be considered a '*trying act*' (Mérei et al., 1987). It is a step towards freedom, i.e. a move in the direction of the integration of the splitted self and the object parts.

2.2. 'An Adoption': Supervision in pairs for advanced trainees

The group leader pair Mona and Fanny brought the theme of a self-experience psychodrama group of adopting parents into the individual supervision. Mona leads a protagonist-centred psychodrama with Mirka, who is described as a 42-

year-old, dark haired woman with an artist-like appearance. Mirka's own child was born in her first marriage, but her new husband, who is 10 years younger than she is, wanted to have his own child. Since Mirka couldn't give birth to another baby because of an operation, they adopted an infant boy. Mirka's concern surprises her groupmates and the leaders as well. She asks if she should go to her son's wedding.

In the self-experience group they created three wedding photographs. Wedding photo A (if she didn't go to her son's wedding): Mirka is not in the picture, only Mirka's first husband with his new wife. Wedding photo B (if she went): Mirka in the middle with her two husbands on her sides. Wedding photo C (psychodrama task): The bride and bridegroom should stand in the middle, decides the group. Mirka arranges the parents in the first row, and the new partners of the parents into the second row. The young couple is placed in the front, so they will be in the middle.

In supervision Mona and Fanny bring up the following questions: "Where is Mirka's place in the play?" Mona and Fanny are first reconstructing the scenes. Mona presents the family members of the protagonist. The people in the family who are not concentrated on are represented by cushions.

While reconstructing the case the reflection on the transference and countertransference is important. Fanny is the more trained one of the two leaders and she is also more dominant. Both of them have an own child and an adopted one. Fanny is still searching for a place among her mother's children. Her mother pampered only her smaller child, a boy, and Fanny was daydreaming about adopting a little girl. In the reflection it becomes clear that Fanny – who was an unwanted child – projects her own situation on Mirka. Playing Mirka's role she could re-live her trauma, and by that she realises that she has no place in her family.

Psychodrama on the countertransference: A psychodrama is formed by the countertransference in supervision. We lined up the female ancestors of Mirka and Fanny (Ancelin-Schützenberger, 1993/2007). Behind Mirka stands her mother who keeps significant distance from her daughter and this distance can be seen between her mother and grandmother as well. Behind Fanny the female ancestors also keep distance from each other, but there is one difference. Fanny placed her grandmother closer to herself than her mother. Her grandmother is a 'cool' woman who represents the female pattern for Fanny.

In the reflection it became clear that there is not only a dynamic between Fanny and her client, but also between Fanny and her colleague Mona. Both leaders' ancestors are members of minorities with difficult backgrounds. But Mona's motherfigures are very loving. Mona was a love child but Fanny was not. Mona can't see clearly while she tries to direct the drama. She identified herself with Fanny on grounds of their minority backgrounds, family situations and similar outlooks, although she was a child much waited for. In the supervision Mona realises that she idealised her leader partner. For her "Fanny was the one who saw the truth". She projected the emancipated, clever woman that she also wanted to be on Fanny.

2.3 'The secret child': Psychodrama training at a telephone helpline service

The following example is drawn from a supervision in a group of telephone helpline service employees, who get trained in psychodrama. Anton is the psychodrama assistant, who leads the group. The participants of the group were reflecting on a client who calls often at the helpline service. Her name is Karina and everybody – except Linda – knows this client very well just by listening to the voice. Anton, the leader of the group, wanted to know more about this client who rang so often, but Linda stopped him several times. Anton felt disturbed by Linda and he became tenser and tenser. In the supervision he wanted to understand better why he reacted so strongly towards Linda.

The supervisor-director focuses on the theme of the conflict between Anton and Linda. Anton is asked by the supervisor what kind of relationship he can associate with this situation. He mentions the employer-employee relation. The supervisor takes this association to form a group-centred drama in which the group members choose the helpline service as the scene of the psychodrama. The group could choose another scene but they are fascinated by the telephone story. They partly reconstruct the situation brought for supervision, but they also use improvisation to enact their fantasies. In their common fantasy Anton is the boss and Linda is his substitute.

Act 1: By means of improvisation the participants present the telephone service with Anton and Linda debating. Then suddenly a telephone booth becomes visible in the corner of the stage. A group member in the role of Karina – her face is covered signalling that her real identity is unknown – calls the telephone service: "I am Karina…"

Anton, who plays himself in the group-centred psychodrama, starts an inner monologue: "Something is disturbing me, I can't listen to my substitute." The supervisor-director of the play steps next to Anton to conduct an interview. Anton thinks of Karina, but also feels disturbed by Linda. At this point the drama becomes protagonist-centred. On the director's advice Anton puts his fantasy with Karina on stage.

Act 2: Karina, the girl from his fantasy, waves to him jumping up and down as if she wants to say something. Anton waves back. Looking at the picture from the outside, from the dramatic mirror, Anton sees a father-like man opposite to the girl, and suddenly he recognises that he brought his first encounter with his first born daughter on stage. Her name is Lilla. He got to know his daughter, who lived abroad, only in her adolescent years, after she called him on the telephone. The supervisor asks him to enact this association.

Act 3: Lilla, the teenager, stands in the booth and calles her father: "It is Lilla here, you know me, don't you? You are my father; I would like to meet you." Anton: "Yes, it's me. I was waiting for your call; I will take you to the most beautiful place in town, the Citadelle. Is tomorrow all right?" This meeting of Anton and her daughter is also enacted on stage. Anton and Lilla are walking around the Citadelle, and they have time to talk to each other. After that scene the supervisor turns the focus back to the theme of the group-centred play.

Act 4: Anton and Linda – boss and substitute – are preparing some tasks they need to handle the following day. Anton listens to Linda peacefully. Linda tells him that she finally also spoke to the mysterious girl, Karina. Anton understands that Linda wants him, the boss and the father figure, to pay attention to her. He smiles and listens to Linda saying: "I think she is not going to call from now on. I told her that everyone knew her from here. She was laughing and said that she sends her regards to everyone. She met someone who is important for her and she will go on her own way."

The group-centred play turned into a protagonist-centred drama when the countertranference feelings of the group leader became the theme. By the change of the method we could follow the main character's unconscious processes. When the countertransference feeling became conscious in Anton, the transference reactions of Linda was not bothering him anymore. This experience at the helpline service with Karina brought up a vision of the 'secret child' in the supervisee.

In the depth of group dynamics the secret that moves the group is hidden: Anton and the female supervisor form the couple that gives life to the 'secret child'. In the group-dynamical process Anton is the central figure of the supervision group. He becomes the main character not only by presenting a case, but also by revealing his countertransference feeling that results in changing the method, i.e. the director changes it to be protagonist-centred.

3. Summary

As a psychodrama director and supervisor I think that in supervision we can attain the most effective work by combination of methods (Szőnyi, 2008). It is always the supervisee's problem that should be in the focal point of the supervision, and it is useful to organise the problem into scenes and actions as early as possible (Avron, 1996). In practice a question arises: With what method can we obtain the proper solution?

I found the answer to this question, and it is the combination of methods. This makes it possible for us to show the group members' inner processes with the help of protagonist-centred drama techniques, and, on the other hand, it teaches us to read and depict the activators of the current group processes with the help of the group-centred drama. The 'reading' of group dynamics is a very important lesson to learn within the training and its practice is necessary because *the problem the group members bring will be reflected, as in a mirror, in the process of the supervision group* (Laveman, 1994). The psychodrama called 'The house' is a good example for this. Here the house symbolised femininity and the desire for seduction as well as the prohibition of it. The symbolics and its meanings appeared also during their own experience play.

Most of the times the psychodrama director candidate is inhibited by his/her own feelings to work with the client. Therefore the reflection on the countertransference is essential in supervision. While working with transferences we can experience the moment of encounter that Stern and his work group (2004) derive from the early harmonisation period of mother and child and which can happen again between the participants of the psychodrama, in my cases between the protagonist and the supervisor. The moment of encounter was most evident between supervisor and supervisee in the drama of 'The secret child'. They waited for the 'secret child' to arrive.

The psychodrama expert, and mostly the supervisor, when directing a psychodrama for supervision, while following his/her forming concept and reconstructing the story of an individual, a couple or a group of people, mixes the methods according to his/her style as a drama director. Then the combinations elaborated by him/her are passed on to the students who will also improve them.

References

Ancelin-Schützenberger, A. (1993/2007). *Aïe, mes aïeux!* Paris: La Méridienne, Desclée de Brouwer.

Anzieu, D., Béjarano, A., Kaës, R., Missenard, A., & Pontalis, J. B. (1973). *Le travail psychanalytique dans les groupes*. Paris: Dunod.

Anzieu, D. (1999). *Le Groupe et l'Inconscient*. Paris: Dunod.

Avron, O. (1996). *La pensée scénique*. Paris, Érès.

Bion, W. R. (1965). *Recherches sur les petits groupes*. Paris: PUF.

Butler, J. (1993): *Bodies that matter. On the discursive limits of "sex"*. London: Routledge.

Erdélyi, I. (1999). The role of the mirror in psychodrama training. In P. Fontaine (Ed.), *Psychodrama Training. A european view* (pp. 173-186). Leuven: FEPTO Publications.

Erdélyi, I. (2010). Mérei Ferenc pszichodrámája (The psychodrama of Ferenc Mérei). In I. Erdélyi, *Mágikus és hétköznapi valóság (Magical and everyday reality* (pp. 172-191*)*. Budapest: Oriold &Tasai.
Hermann, I. (1949). The Giant Mother, the phallic Mother, Obscenity. *Psychonanalytical Review*, 36, 302-306.
Holmes, P. (1992). *The inner world outside: Object relations theory and psychodrama*. London, New York: Routledge.
Kernberg, O. (1979). *Les troubles limites de la personnalité*. Paris: Privat.
Klein, M. (1955/1986). A Study of Envy and Gratitude. In J. Mitchell (Ed.), *The selected Melanie Klein*. London: Hogarth Press.
Laveman, L. (1994). Resemblance and difference between supervision and psychotherapy. *The Clinical Supervisor*, 12 (2).
Mérei, F., Ajkay, K., Dobos, E., & Erdélyi. I. (1987). *A pszichodráma önismereti és terápiás alkalmazása (Applying psychodrama for self-experience and therapy)*. Budapest: Akadémiai Kiadó.
Moreno, J.L. (1965). *Psychothérapie et psychodrame*. Paris: PUF.
Moscovici, S. (1972). *Psychologie des minorités actives*. Paris: PUF.
Nève – Hanquet, Ch., & Van der Borght, Ch. (1999). Doubled chairs in supervision. In P. Fontaine (Ed.), *Psychodrama Training. A european view* (pp. 269-274). Leuven: FEPTO Publications.
Stern, D. N. (2004). *The Present Moment in Psychotherapy and Everyday-Life*. London: Norton & Company.
Szőnyi, G. (2008). A kezelési kombináció kérdései (Problems of method combination in treatment). In G. Szőnyi, & Füredi, J. (Eds.), *A pszichoterápia tankönyve, 2. kiadás* (*Study book of psychotherapy, 2nd ed.)*. Budapest: Medicina.
Winnicott, D. W. (1971). *Playing and reality*. London: Tavistock Publication.

'Exchanging Ideas' on stage and developing a professional identity – practical applications in supervision

Hilde Gött

For many years now I have been guiding trainees through their individual and group supervision. My aim in supervision is to use the work on stage to develop confidence of the students in their professional roles and to help them to find their identity as a psychodramatist.

In order to illustrate my basic ideas and assumptions of the subject, I will first outline my understanding of supervision (Pühl, 1994; Buer & Siller, 2004; Gellert & Nowak, 2007) and explain the settings of group supervision which I offer to my trainees. I will then describe two exercises, which I found very helpful in supervision. The first one will focus on an 'exchange of ideas' among trainees; the second one will put emphasis on reflecting different roles of a therapist and how to integrate them.

1. Supervision in psychodrama training

Psychodrama training should be open to diverse professional fields such as pedagogy, psychology, theatre, and organizational development and also open to fields where it has not been previously applied. Psychodrama training is a long-term process that requires patience, courage and initiative. When trainees later implement these methods in their own practice they usually work alone and despite all their uncertainties they have to react spontaneously and effectively in all unforeseen situations. Supervision should facilitate and reflect this transition to the psychodrama practice and should allow especially the less experienced psychodramatists to build and strengthen their self-confidence.

Supervision is a very important part of the training programme in the upper level. When the trainees apply psychodrama in their practical work they have to have 90 hours of supervision (minimum 10 hours in the individual setting and a minimum of 60 hours in a group setting).

Group supervision has the format of a psychodrama seminar, which is held in the weekend (18 training hours of 60 minutes). The supervision begins with an introductory round, where each supervisee describes his or her current situation and formulates any questions, expectations or wishes they may have. The weekend then proceeds in a way in which the topics and steps of the processes are constantly re-formulated. Just as in other psychodrama-group sessions, supervision includes enactment, role feedback, scenic representation and the 'surplus reality' of the stage. Collaborative work among peers of trainees and learning from each other as professionals is very important as well.

The principal aim of supervision is to improve the quality of professional behavior ('German Society for Supervision'/ Deutsche Gesellschaft für Supervision) (Buer, 1999, 2004).

Supervision should help trainees to achieve a greater degree of competence in their professional work with clients. Furthermore trainees should get support to develop and integrate different roles in therapy and to find their identity as a psychotherapist. In this training process the supervisor has the role of an experienced critical expert and a sensitive teacher. Supervision can focus either on the work with a client or on more general theoretical, methodological or philosophical issues. Both aspects should be addressed in supervision.

In this article I will describe the process and examples of two methods which I found very helpful in group supervision. One is called 'the exchange of ideas' on stage. The other one will focus on the integration of different roles of a therapist.

2. 'Exchange of ideas' in the supervision process

The 'exchange of ideas'[1] on stage requires an anxiety-free and forgiving atmosphere in the group. Sociometric work at the beginning can help the trainees to learn respect for each other and for this teaching model. The model follows certain steps, which will be described and illustrated by an example how it can be applied in group supervision. The roles in this model are: supervisee A = protagonist; group member B = client; group members C, D, E, F = alter ego.

Enactment of a scene, which entails the question of the supervisee: A situation or interaction of psychotherapeutic work is performed through psychodrama where the usual role giving and role taking crystallizes the problem on stage. This problem can relate to either a group or an individual setting. When the interaction has been played through the supervisor stops or "freezes" the scene.

Role feedback: By listening to the role feedback and suggestions from the players on stage new perspectives can emerge, which add new insights to the initial description of the problem. On the basis of that a new or more appropriate

[1] The 'exchange of ideas' on the psychodramatic stage follows a process, which I learned from Ildikó Mävers, who guided me through my own supervision training.

diagnosis of the problem can be formulated. In a next step different possible interventions can be explored

'Exchange of ideas' as a training exercise: In the following step the first sequence will be replayed in the same constellation as before. One of the other supervisees (F, G, H etc.) will now lead the group and try how the therapist can intervene in a helpful way. It is essential that supervisee A takes the role of his or her own client and plays it as the antagonist. All the others replay the roles they had in the first scene. Eventually the supervisor will interrupt and freeze the scene and request new feedback, especially that which reflects the role of supervisee A as client and of the therapist about his or her intervention. This scene can be repeated with different members of the group in the role as a therapist. At the same time the supervisee has different opportunities to explore the role of the client and to understand better which intervention can bring new insights and progress to the client. At the end of this exploration the supervisee A finishes the work on stage in his or her own role as supervisee. An example of group supervision with this 'exchange of ideas' on the psychodramatic stage can illustrate how this model works.

'Exchange of ideas' – illustrated by an example in supervision: Maria is a psychotherapist in a psychosomatic clinic. In her group supervision she introduces a patient, who is one of 14 patients in a therapeutic group, which meets twice a week for 90 minutes. After having spent four weeks at the clinic Georg revealed that his wife left him two weeks before he entered the clinic. Maria was not sure how to deal with that situation of Georg in the group.

At the beginning in supervision Maria presents the situation of the patient Georg as he explained it to her in the therapy group. Then she describes her therapeutic intervention and how she was dealing with the problem of the patient in the group. As described above in the first step the group enacts the scene of the therapeutic work. Maria takes on the role of the patient Georg.

In the next step the supervisor freezes the scene and requests role feedback from the participants on the stage, especially from the person taking the role of Georg. Then the 'exchange of ideas' can begin: Maria stays in the role of the patient. One by one the others supervisees take on the role of the therapist in order to explore how they would handle the situation.

First attempt: The first attempt is done by Andreas who takes on the role of the therapist. He asks Maria in her role of the patient Georg how life has changed after the separation from his wife and what he intends to do in future. After this intervention the supervisor stops the play and asks Maria for feedback in her role of Georg. After that the supervisor asks Andreas about his intentions in the role of group therapist. Andreas wants to look at the future together with his patient and to support his next steps in his life.

Second attempt: This time Anna takes the role as therapist and asks the patient to reflect on the time when he left his home and went to the clinic after his wife told him about the separation. This time the supervisor also asks for a role feedback after this intervention. In her role as therapist Anna wants to provide the patient Georg with space to contemplate his feelings about the separation and allow him to mourn for the relationship.

Third attempt: This time Michael takes the role of the therapist and tries a new intervention. He encourages the patient (Maria in the role of Georg) to take a close look at his current situation. He asks about the situation in the clinic, about Georg's separation from his home and his uncertainty about the future. Michael proposes that Georg could accept the distance to his family, his wife and his home and acknowledge his problematic family contact in the past. In the role feedback on stage Maria tells that she had gained insight from playing the role of the patient Georg in his current situation and now intends to talk with the family again.

This example illustrates how three supervisees were able to experience and respond to the same therapeutic situation. In the role of a therapist they were able to form different hypotheses and they could test their interventions. Finally they could explore and determine which of these interventions is more helpful than the other. By doing this the 'exchange of ideas' can be a bridge between the 'surplus reality' in a learning situations during supervision and the real psychotherapeutic work.

3. Professional identity and the integration of different roles

The supervisor should encourage and help the supervisee to gain confidence in their professional development. Trainees have to find their identity as a psychodramatist and therefore they have to integrate different roles. The concept of different roles, which has been described by Kellermann, provides a structure for reflection. He distinguishes different roles by referring to Moreno:

"All psychodramatists enact a few specific and sometimes overlapping roles. Moreno (1972) described theses as the roles of producer, of therapist/counselor and of analyst. I have revised the meaning of these roles and have added a fourth role which I feel is intrinsic to psychodrama, namely that of the group leader. ...these roles and their function, skills and ideals which together comprise the professional demands on the psychodramatist." (Kellermann, 1992, p. 46)

As *group leader* the psychodramatist has to be aware of the group, the members, the topics, the conflicts and the development of each individual participant. He or she will look closely at the actual situation and themes of the group as a whole, and also at situations where topics occur simultaneously. He or she is responsible for the atmosphere in the group, the cohesion, the confrontations between the members and the resulting integration of the group.

As *therapist* the psychodramatist concentrates on each individual client as protagonist. He or she is responsible for the development of the topic, for the confidence and actions of the protagonists and their integration in the group processes.

As *producer*, the psychodramatist is responsible for staging and directing the scenes, the explanation and implementation of the roles, the art and style of the drama on stage and not least for the aesthetics of the production.

As *analyst*, the psychodramatist is responsible for analysis of the group, the situation, the protagonists, the diagnoses and the therapeutic concept as a theoretical construct.

Each role is equally important. Strengths and weaknesses of the supervisees can be evaluated by looking at the competencies, which are related to these different roles. In supervision the supervisor explains the four roles and the supervisee is invited to asses himself by playing each of the four roles. In doing so the protagonist moves from one chair to another. Afterwards the supervisor can put a fifth chair in front of the others for the personal view of the 'psychodramatist'. After the self-evaluation the supervisee sits on the fifth chair to connect the four roles with the own personality.

Integration of roles – illustrated by an example in supervision: Alexander is in the advanced course of the psychodrama training program and although he conducts successful protagonist work in his training group he does not like to invite patients to work on stage. We enact a situation with his patients in individual therapy. The supervisee knows about methods and techniques in psychodrama, but he hesitates to invite the patient to work on stage.

The supervisor suggests looking at his role repertoire as a psychodramatist and he puts four chairs on the stage and defines the roles as a leader of the group, therapist, producer and analyst. Alexander explores one role after the other by sitting on the chair and describing his experiences and confidence in these different roles:

Alexander starts with the 4th chair, the role of the analyst. He says that he is a good listener, can easily diagnose, can associate well and knows how he should handle processes with the patient. In the 2nd chair, the therapist, he tells that he knows how to work with the patient and how to encourage him to talk about his thoughts and feelings. He feels very confident in this role.

In the 1st chair in the role of a group leader Alexander talks about his good experiences in a group, which he conducted previously. In the 3rd chair, the role of the producer/ director, he talks about his difficulties to work on stage with the patient. Simply the fact to be on stage as a director makes it difficult for him. He talks about the differences to be on stage as a protagonist or as a director.

At the end of this exercise I as the supervisor put a fifth chair for his 'personal role' in front of the other chairs and I ask him to think about why he does not invite his patients to work on stage. On the chair Alexander says that he is an inhibited person and it is easier for him to remain in the background. At the same time he remembers being a protagonist and tells from his experience as a protagonist how it feels well to work on stage and to follow his impulses. He remembers that it was the most important experience and motivation to start psychodrama training. Being aware of his own feelings as a protagonist we return in the closing scene on stage to his treatment room in the situation with his patient. When the patient begins to talk about his topics, Alexander stands up and invites him to come on the stage in order to create a scene.

4. Summary

To summarize I would like to point out that the trainees enjoy these two exercises because they produce different ideas to resolve the problem, they easily develop their own style and feel encouraged to admit mistakes they make and not to hide them. They can work in an anxiety-free atmosphere, forgive mistakes and correct them. The two exercises I use in the supervision of trainees are especially useful to…

- raise practical competence
- give trainees the courage to experiment and learn in an anxiety-free atmosphere
- support the transition to work in one's own psychodramatic practice
- enhance spontaneity, flexibility and tolerance
- self-evaluate and to give professional feedback
- gain a professional identity as a psychodramatist
- build a network of psychodramatist colleagues

Some of my supervisees have remained longer than usual in supervision groups in order to take advantage of the diversity of learning and methods in the groups and to advance and develop their personal and professional identity.

References

Buer, F. (1999). *Lehrbuch der Supervision. Schriften aus der Deutschen Gesellschaft für Supervision e.V.* Münster: Votum Verlag.

Buer, F. (2004). *Praxis der psychodramatischen Supervision. Ein Handbuch.* Wiesbaden: VS-Verlag für Sozialwissenschaften.

Buer, F. & Siller, G. (2004). *Die flexible Supervision. Herausforderungen – Konzepte – Perspektiven. Eine kritische Bestandsaufnahme.* Wiesbaden: VS-Verlag für Sozialwissenschaften.

Gellert, M., & Nowak, C. (2007). *Teamarbeit – Teamentwicklung – Teamberatung.* Meezen: Verlag Christa Limmers.

Kellermann, P. F. (1992). *Focus on psychodrama.* London: Jessica Kingsley Publishers.

Pühl, H. (1994). *Handbuch der Supervision.* Berlin: Edition Marhold im Wissenschaftsverlag Volker Spiess.

IV. Training and Research in Supervision

Training and research in supervision – an introduction

Hannes Krall

In the psychodrama community there is a broad consensus about the importance of supervision for the professional development of trainees, which is emphasised by the european training standards (FEPTO, n.d.). As the previous chapters in this book illustrate, supervision in psychodrama training has created divers practices and traditions over time. From the knowledge about psychodrama training in different institutes and from this book one can assume, that supervision is well based on different theoretical and philosophical concepts, and it also has a very rich basis in terms of different experiences and practices. On the other hand there is still not enough empirical research about the effects of supervision on trainees, their learning and development in their psychodrama practices with individuals and groups. Therefore, the following section in this book will look at examples in supervision which address training, evaluation and research.

1. Research on supervision

It is surprising that in research supervision does not get much attention (Schigl, 2008, p. 39) even though there is a growing demand for more research (Krall, Mikula & Jansche, 2008; Buer, 2008; Galdynski & Kuehl, 2009). This is especially true for supervision in psychotherapy training. Even though there have been studies conducted for quite a long time (e.g. Ronnestad & Skovholt, 1993, 1997; Ronnestad, Orlinsky, Parks, & Davis, 1997; Wheeler & Richards, 2007) supervision in psychotherapy training is still a neglected field of research.

So far, research studies on effectiveness put emphasis on the complexity of different aspects which cause positive results in supervision (Worthen & McNeill, 1996). Some studies are also analysing and reporting negative effects (Ruskin, 1994; Gray et al., 2001). However, it is important to take into consideration that the perception of supervisees and supervisor of processes and outcomes can differ significantly (Reichelt & Skjerve, 2002).

Broad consensus can be found about the fact, that the quality of the relationship between supervisee and supervisor is a crucial factor. On the other hand unresolved conflicts in the relationship are most often the reason for negative experiences in supervision (Nelson & Friedländer, 2001).

Of course, a good relationship can only contribute to a successful supervision, if the applied methods and techniques are appropriate. Active methods which use enactment or sociometric constellations are described and promoted as helpful (e.g. Williams 1995; Buer, 2001; Krall & Schulze, 2004; Chesner, 2008; Krall, 2009). If these methods are used in supervision it should get more attention in research as well.

2. Supervision as practitioner research

In supervision trainees are used to reflect upon their case work by listening, observing, documenting and analysing. Furthermore this reflection is deepened or broadened by the reflection with a supervisor and peer-trainees in the supervision group. They bring another perspective into the reflection, ask additional questions in order to broaden or challenge the reflection of the trainee and to give feedback whether the interventions seem to be appropriate or not. This reflective work includes the diagnosis and the progresses of clients, the interactive processes in the group – between the director, the group and the individual client –, the reflection of the applied techniques and methods and the ongoing self-evaluation of the trainee. This kind of reflective practice in supervision usually is the baseline for every psychodrama trainee.

Since reflection in supervision can take place not only with action methods, but also with research tools and procedures to evaluate the work of the trainee, supervision follows a model of a collaborative practitioner research in order to co-create new practical knowledge, which aims at good quality in psychodrama. Supervision in this context is basically a systematic collaborative endeavor to learn from practical experiences and therefore it can be regarded as a kind of practitioner research (Krall, Mikula, & Jansche, 2008). In this way supervision can serve as a bridge between psychodrama practice, professional standards and research.

In supervision the trainee should be encouraged to take on and to strengthen the role as a practitioner researcher. The TRAIN Project (Towards Research Applied in International Networks of Trainees) for example is proposing to implement systematic reflection on the practices of trainees in psychodrama. Furthermore it stimulates international exchange among trainees to share their experiences and outcomes in research (Krall, Fürst, & Doganer 2009; Krall & Fürst, 2011). The trainee gets acquainted with research tools and learns to see research as helpful for psychodrama practice rather than perceiving it as something which is alien to the practitioner (Krall & Fürst, 2009).

In this context supervision of trainees will play a crucial role to support trainees to do their own systematic reflection by applying research tools and procedures. And vice versa research of the trainees on their psychodrama practices can enrich the process of reflection in supervision.

Trainees as reflective practitioners (Schön, 1983) need a capacity to inform, to question, to develop their own practice by referring to research. It is inevitable to become acquainted with basic research designs in counselling, psychotherapy and other professions to articulate relevant research questions, to gather significant data, to analyse and interpret them properly and to come to evidence based conclusions.

3. Supervision in psychodrama training

Most psychodrama training institutes apply supervision in different phases and contexts. In one part oft the training the trainees get supervision on their practices as director or co-director among peers in their own training group. In other phases of the training the trainees might get supervision during their practicum when they are in their role as co-workers in a professional field like education, psychology or health services. And of course, all trainees get supervision when they are in the role of conducting psychodrama groups on their own.

If supervision plays an important role in learning and professional development of trainees, one consequently hast to discuss more deeply the theoretical and empirical foundation of supervision practices. At present, how supervision is done seems to depend very much on the personal experiences of the supervisors and the traditions in different training instituts. A brief look at the different ways of doing supervision as they are described in this book demonstrates, that supervision can be conducted by applying psychodramatic methods and techniques (Teszáry; Fürst & Krall), but also by doing it differently like in a group analytic way (Tsegos & Karapostoli). It can be conducted in a format similar to a psychodrama group (Gött), but it is also possible to do it according to a unique procedure of reflection (Boria). The chapters about supervision in this book refer to the philosophy and theory of Moreno, but they include also other approaches like the psychoanayltical concepts of transference and countertransference (Erdélyi) or Jungian psychoanalytical approaches (Gasseau & Perrotta), systemic-constructivist concepts (Dudler & Weiß) or client centered theory (Apter).

It is interesting to see all this different approaches and philosophies of doing supervision in psychodrama training. Although there is a considerable variety of different procedures, methods and techniques in supervision, very rarely they have been systematically scrutinized upon their effects on the learning of the trainees. What is eventually the outcome of learning in supervision? What can be transferred in the ongoing practice of the trainee? What are helpful and hindering aspects of supervision? Does supervision fail and why? How is the working alliance in supervision strengthened and how important are personal characteristics of the supervisor and the trainee for a successful learning experience in supervision? Answers could be found by enhanced sharing of experiences amongst supervisors, discussions about theory, practice and training in supervision, and by a systematic research on supervision in psychodrama training. So far, however, there are just a few studies mostly associated with single approaches in supervision. Nevertheless it is an opportunity to start with and to put more emphasis on training and research. Therefore in the following chapter training, evaluation and research in supervision will be focused on.

Anna Chesner introduces a one year supervisors training course, which is based on psychodramatic and related action methods. In this chapter the philosophy of the course, its structure and some specific action based methods for supervision are described.

Pierre Fontaine reports about his experiences to introduce research in psychodrama training. He wants to encourage trainees to do bottom-up research in their practice. In his chapter he describes the concept of research in psychodrama training, first steps with trainees and some practical examples of practitioner research projects.

In a similar attempt Gabriela Moita and António Roma-Torres discuss the possibility to integrate research in the training programme of psychodramatist. Supervision is seen as an ideal phase for the acquisition of research skills during psychodrama training. In this chapter they present their experiences of teaching research methodologies and monitoring of project development during supervision in psychodrama.

Examples of how to do supervision in a group setting and how to evaluate the outcome is given by Ioannis K. Tsegos and Natassa Karapostoli. They describe a specific model of supervision, which provides a structured procedure to focus on the relationship between a therapeutic group and its supervising group. Research findings on this model are presented and discussed.

Jutta Fürst and Hannes Krall introduce their supervision concept, structure and setting in psychodrama training. Supervision practice will be illustrated by practical examples and discussed by theoretical considerations. Finally the focus will be on the results of a pilot study to evaluate outcomes and helpful factors in supervision.

References

Buer, F. (Ed.) (2001). *Praxis der psychodramatischen Supervision*. Opladen: Leske + Budrich.

Buer, F. (2008). Erfahrung – Wissenschaft – Philosophie. Drei Wissenssorten zur Konzipierung von Supervision und Coaching. In H. Krall, E. Mikula, & W. Jansche (Eds.), *Supervision und Coaching. Praxisforschung und Beratung im Sozial- und Bildungsbereich* (pp. 223-238). Wiesbaden: VS Verlag für Sozialwissenschaften.

Chesner, A. (2008). Psychodrama: A Passion for Action and Non-Action in Supervision. In R. Shohet (Ed.), *Passionate Supervision* (pp. 132-149). London and Philadelphia: Jessica Kingsley Publishers.

FEPTO (Federal European Psychodrama Training Organziation) (n.d.). Retrieved from http://www.fepto.eu/web/en/Minimal_Training_Standards, (January 4, 2011).

Fürst, J. & Krall, H. (2011). TRAIN -Towards Research Applied in International Networks of Trainees. In Book of Abstracts of the 42nd Annual Meeting of the Society for Psychotherapy Research (SPR), June 29 to July 2, 2011, Bern, Switzerland (p. 72).

Gladynski, K. &. Kühl, S. (Eds.) (2009). *Black-Box Beratung? Empirische Studien zu Coaching und Supervision.* Wiesbaden: VS Verlag für Sozialwissenschaften.

Gray, L. A., Ladany, N., & Walker, J. (2001). Psychotherapy Trainees' Experience of Counterproductive Events in Supervision. *Journal of Counseling Psychology, 48 (4),* 371-383.

Krall, H., Fürst, J., & Doganer, I. (2009). *TRAIN - Towards Research in an International Network of Trainees.* In Minutes of the Catania FEPTO RC Meeting October 16-18, 2009.

Krall, H. & Fürst, J. (2009*). Research in Psychodrama Training.* Paper presented at the 17.th Congress International Association for Group Psychotherapy and Group Processes (I.A.G.P.), Rome August 24-29.

Krall, H. & Schulze, S. (2004). Psychodrama in der Supervision und im Coaching. In J. Fürst, K. Ottomeyer, & H. Pruckner (Eds.), *Psychodrama-Therapie. Ein Handbuch* (pp. 412-423). Wien: Verlag Facultas.

Krall, H., Mikula, E. & Jansche, W. (Eds.) (2008). *Supervision und Coaching. Praxisforschung und Beratung im Sozial- und Bildungsbereich.* Wiesbaden: VS Verlag für Sozialwissenschaften

Krall, H. (2009). Sprache und Szenische Arbeit in der Supervision. *Supervision. Mensch, Arbeit, Organisation,* 2, 19-24.

Nelson, M. L. & Friedlander, M. L. (2001). A Close Look at Conflictual Supervisory Relationsships: The Trainee's Perspective. *Journal of Counseling Psychology 48 (4)*, 384-395.

Reichelt, S. & Skjerve, J. (2002). Correspondence between Supervisors and Trainees in their Perception of Supervision Events. *Journal of Clinical Psychology, 58 (7)*, 759-772.

Ronnestad, M. H. & Skovholt, T. M. (1993). Supervision of Beginning and Advanced Graduate Students of Counseling and Psychotherapy. *Journal of Counseling and Development 71*, 396-405.

Ronnestad, M. H. & Skovholt, T. M. (1997). Berufliche Entwicklung und Supervision von Psychotherapeuten. *Psychotherapeut (5)*, 299-306.

Ronnestad, M. H., Orlinsky, D. E., Parks, B. K., Davis, J.D., Society for Psychotherapy Research (SPR) Collaborative Research Network (1997). Supervisor of Psychotherapy: Mapping Experience Level and Supervisory Confidence. *European Psychologist 2 (3)*, 191-201.

Ruskin, R. (1994). When Supervision May Fail: Difficulties and Impasses. In S. E. Greben & R. Ruskin (Ed.) *Clinical Perspectives on Psychotherapy Supervision* (pp. 213-261). Washington: American Psychiatric Press.

Schön, D. (1983). *The Reflective Practitioner.* London: Temple Smith.

Schigl, B. (2008). Supervision: Ergebnisse aus der Forschung oder was brauchen ForscherInnen und PraktikerInnen voneinander? In H. Krall, E. Mikula & W. Jansche (Ed.), *Supervision und Coaching. Praxisforschung und Beratung im Sozial- und Bildungsbereich* (pp. 39-52). Wiesbaden: VS Verlag für Sozialwissenschaften.

Wheeler, S., & Richards, K. (2007). *The impact of clinical supervision on counsellors and therapists, their practice and their clients: a systematic review of the literature.* Lutterworth: British Association for Counselling and Psychotherapy.

Williams, A. (1995). *Visual and Active Supervision.* New York: W.W. Norton & Company.

Worthen, V. u. McNeill, B. W. (1996). A Phenomenological Investigation of 'Good' Supervision Events. *Journal of Counseling Psychology 43 (1)*, 25-34.

Psychodrama based supervision training

Anna Chesner

Over the past ten years I have been involved in the development and delivery of a one year supervisors training course: Creative Approaches to Supervision. It has its roots in a collaboration between Dr Sue Jennings and myself through Rowan Studio and the Institute of Action Methods, it now runs under the umbrella of the London Centre for Psychodrama. The course is unusual in that it is firmly based in Creative Methods, whilst being a generic as opposed to specialist supervision training. In practice this means that we are training people to supervise across disciplines, whilst using the language of psychodrama and action methods as a core element in terms of theory, teaching style and supervision techniques.

As a direct consequence of this approach we typically attract practitioners from a variety of professional disciplines. In recent years we have had psychodrama psychotherapists, dramatherapists, art psychotherapists, dance movement therapists, play therapists, integrative child psychotherapists, systemic family therapists, psychoanalytic psychotherapists, counsellors, social workers, coaches and those involved in church ministry and training. Having such diversity within the training group fosters dialogue, encounter, and a high degree of peer level learning. This is central to the philosophy of the course.

Some of this diversity of approach is reflected in the core teaching team. The two core trainers (Anna Chesner & Lia Zografou) between us draw on our background in dramatherapy, psychodrama, group analysis, gestalt, organisational consultancy, and playback theatre.

1. Status of the course

The training is at postgraduate post-professional diploma level, comprising a minimum of 130 taught hours, which are delivered in London, at the Maudsley Hospital, London over six weekends.

The course was recognised firstly by the British Association of Dramatherapists (BADTh), which for many years has kept a register of qualified supervisors who are available to their membership for supervision whilst training. BADth also has a system whereby full membership of the association is dependent on practitioners evidencing 40 hours of post-qualification supervision. As such they have been exemplary in giving supervision a high profile within the profession. I have written elsewhere (Chesner, 1999) about the historical background to this; a development driven largely in the late 1980s by Dorothy Langley, a doubly trained dramatherapist and psychodrama psychotherapist.The course has for the past four years been accredited by the British Psychodrama Association (BPA). It is one of two trainings accredited by them; the other being a person-centred training in supervision founded by Jenny Biancardi.

The double recognition and accreditation of the London Centre training in Creative Approaches to Supervision means that it has a standing with professional organisations which are based within the Health Professions Council (HPC) and the United Kingdom Council for Psychotherapy (UKCP), giving it a broad range of recognition across disciplines and professions which may be allied or seeking membership of either of these umbrella organisations. To date trainees have attended from across the UK, Greece and Ireland.

While the factors relating to external validation continually change, the philosophy of being responsive to the validation needs of different professions has ensured a wide spectrum of different therapists – which we see as a definite asset to the training. In an increasingly regulated culture external validation and quality control matter.

2. Entry to the training

When is someone ready to train as and practice as a supervisor? We have a minimum requirement of three years post-qualifying practice. Some applicants have been supervising for many years and do the training to gain a formal qualification (increasingly important for eligibility for some jobs) or to learn new and creative approaches to supervision. Others are new to the role of supervisor and may be being asked to supervise trainee therapists on placement at their work place. In the current climate there is a tendency for untrained workers to take quite high levels of responsibility in the workplace, and in these circumstances relatively junior trained practitioners may be asked to supervise them.

The amount of practice and clinical experience a person accumulates in three years varies enormously and this is assessed through the formal application and at interview.

Applicants are required to send a CV, evidence their professional memberships and professional insurance. In addition they provide two referees, one of

whom is their current supervisor. These are approached for a confidential reference as to suitability for the course at this time and any concerns.
The interview process tends to be quite collaborative, an opportunity to tease out the degree of readiness for the training, to look realistically at the commitment required and the approach of the course, and to give attention to any anxieties which arise. There is a paradoxical status to the trainee supervisor role. On the one hand there is a requirement for a good professional practice base (i.e. 'I feel ready for the next step professionally'), on the other hand it is a training role. ('I am at the beginning, and I haven't been at the beginning since the start of my professional training') It is not uncommon for applicants to feel vulnerable about their clinical practice at the outset. The prospect of being supervised on their clinical practice as part of the training process within the group can feel exposing. Another source of anxiety comes from previous training traumas or supervision conflicts. It is helpful for these anxieties to be acknowledged authentically at interview. On occasion the training itself has been reparative for participants who come with this kind of unfinished business.

3. Structure of the training

The course content is divided into 4 modules: Supervisory roles and supervisory relationships (two weekends), Practice Issues (two weekends), Specialist Perspectives (one double weekend), Supervision on Supervision (one weekend and 8 individual reflective practice sessions).

The supervision on supervision (reflective practice) sessions run concurrently with the other modules, and the apparent sequence of the modular structure belies the fact that specialist perspectives and practice issues are a focus from early on in the course. A little more about the content of the modules:

Supervisory roles and supervisory relationships: setting the scene
The first module covers the following content:

- Orientation to the course: setting a group learning contract; going through the course document and establishing a timetable of milestones.
- Getting an overview of the field: Exploration of supervision settings and perspectives; different models and experiences of supervision.
- Exploration of the wider systemic perspective on supervision, including the network of stakeholders from client to the general public.
- Settings and frames, and their impact. One to one, team and group supervision: relationship issues informed by setting.
- Theories of Supervision: Supervisor roles

- Focus of intervention: seven eyed model of Hawkins and Shohet
- Practical considerations (information gathering, contract making, working with parallel process, recording)
- Ethical dimensions – an exploration of different codes of practice and questions of accountability
- Action Techniques (see below)

By the second weekend (a month into the training) trainees are expected to be actively supervising, so that the taught elements of the course are integrated into practice.

Practice issues: doing the doing
- The second module links theory to practice
- Qualities and skills of the developing supervisor
- Exploration of theoretical diversity: case exploration from contrasting theoretical perspectives
- There is a focus on personal values and professionalism in relation to regular and crisis supervision
- Boundary issues with therapy
- Accountability, legal and anti-discriminatory practice issues

Specialist perspectives: details and specifics
- The third module Specialist perspectives in terms of particular client groups (eating disorders, children, forensic, dual diagnosis, addiction, personality disorder,)
- Specialist perspectives in terms of contexts (medical, community, social services, private practice, education, forensic)
- Advertising and networking. Establishing a practice, setting a fee scale, self and peer evaluation; course review.

Supervision on supervision:
- In the fourth module we devote one weekend to using the training group to reflect in action on how the supervisory (as opposed to the clinical practice) is going. This takes place quite early in the process and comprises the third weekend of the training.
- We also use this weekend to pay particular attention to issues arising around supervising teams and groups.
- In addition, there is a requirement throughout the course to attend at least eight one to one sessions of supervision on supervision, with a formally qualified

supervisor who has an affinity with the use of creative interventions in supervision practice.

4. Theoretical base

The teaching style is in action. Each time we meet there is a session on theory of supervision, which we explore using sociodramatic, concretising or improvising techniques. There are three main theoretical sources:

- Williams' role based approach to supervision (Williams, 1995). Williams offers clear perspectives on the importance of a wide and consciously used role repertoire for the supervisor. He specifically gives examples of teacher, consultant, facilitator and evaluator roles. To these we have added a fifth, the administrator role. Trainee supervisors are encouraged to develop an awareness of which of these are given priority in their practice and why.
- Hawkins and Shohet 7-eyed supervisor model (Hawkins & Shohet 2000). This provides a self-reflective frame for the supervisor, in which he or she has the option to focus the session through specific lenses, and to reflect on whether there are certain lenses which are used as a cultural conserve, whilst others are relegated to the status of underdeveloped or underused roles, or even blind spots. In brief the eyes or lenses relate to focusing on a) the story or content of the session; b) strategies or interventions used; c) the therapeutic alliance and client transference; d) tele and countertransference issues; e) parallel process and the supervisory relationship; f) the supervisor's countertransference reactions in the here and now of the supervision session; and g) the wider (social, organizational and transpersonal) perspective
- The above are our two core texts on the training, and we supplement them with a 'Supervisory question' model that has been developed out of the London Centre role analysis approach to psychodrama. The supervisory question equates to the London Centre emphasis on clear "Contracting" in psychodrama psychotherapy. Supervisors are encouraged to tease out and develop one or more supervisory questions in a session, so that the exploratory and reflective activities of the session relate directly to a shared intentional focus. Many of us have experienced supervision which operates like a quality control checklist – based on a monitoring role of the supervisor. While this approach may reveal the basics of who the supervisee is seeing, how they are attending, how many sessions they have done and whether anything majorly disturbing is taking place, it does not foster in-depth exploration of the relationship or the work being done. Another model of supervision is a looser approach of joint reflection and a free- floating discussion around the work. This has the advantage of

permitting some depth, but may not address the most salient concerns of the supervisee and may end up habitually expressing the academic or clinical interests of the supervisor. It may also neglect the story of what has happened over process questions. The 'Supervisory question' model ensures that the supervisor is listening for a key focus in a session or part of session, and makes it explicit. Supervisor and supervisee focus on this together until moving on to a next issue.

Example of finding the Supervisory Question:

Supervisee: Oh, I'm so tired...I really don't feel like being here today. It's something to do with what's going on my group...no-one turned up again last week!

Supervisor: Why do you think that is?

Supervisee: I don't know, it's been a while since the group attendance has been regular...I was on holiday, then we had the new member join...oh and now I think about it, last time he came I didn't do a great piece of work, and he said he was overwhelmed and psychodrama wasn't for him...and he hasn't been back since...

Supervisor: (teasing out the focus): It seems that it might be helpful to look at that piece of work and what you might learn from it; or is this more about thinking about the disrupted experience of this group as a whole and its impact?

Supervisee: It's both those things and how that has affected attendance.

Supervisor: So, let's make that our focus: how the recent experience of the group and the "not great piece of work" with your new member may be affecting attendance in your group.

In addition, if there are other theoretical learning needs, e.g. a review of psychological mechanisms such as transference or projective identification we include experiential theoretical input as and when needed in the group.

Towards the end of the course trainees are asked to make a brief written analysis of a supervision session they have facilitated in terms of the following criteria, designed to help them integrate the key elements of theory with their developing practice: supervisory question/ issues; verbatim/ action interventions; which eye used; predominant supervisor role; creative action method used.

5. Teaching style

As well as the session on supervision theory as outlined above each weekend there is a demonstration and practice session of a creative action method as a supervision tool. One of the tutors demonstrates with a member of the group as supervisee and the rest of the group in reflective witness role. The group then reflects together after the demonstration, naming the supervisory question heard, the processes and techniques witnessed, and reflecting on the impact and timing of the action approaches.

This is an essential prerequisite before going into practice triads/ small groups. One of the challenges to the trainee supervisor is to learn to focus on the process rather than the content of the supervision session. As therapists there is a tendency to identify with the supervisee-client subsystem, and to overly engage with the client or situation presented. The danger of this is that it produces a "Why don't you do it like me?" style of supervisor, in covert competition with the supervisee. So the reflective process as a group encourages a meta-level perspective on the interaction, that acknowledges the supervisee-client or supervisee-workplace subsystem, but also allows for focus on the supervisee-supervisor subsystem, and the supervisor's process in terms of decisions made during the supervision session.

These group reflections inform the subsequent triad or small group practice sessions throughout the weekend.

5.1 Triad practice

Triad and small group practice form a key part of the learning, providing an opportunity to develop the role of supervisor and link theory to practice within a supportive and reflective environment.

We facilitate a warm up to the roles and tasks of supervisor, supervisee, and observer before embarking on this work. The supervisee's task is to bring an authentic and current work-based issue that is not in itself supervision, generally something clinical, collegial or organizational.

The supervisor's task is to manage the space and time of the session, identify and tease out the supervisory question/focus, practice a new creative action technique or incorporate it into already familiar techniques.

The observer's focus is primarily on the supervisor rather than the supervisee, although the quality of their relationship is significant. He or she may concentrate specifically on identifying the supervisory question, the eyes/lenses used in exploring it and the supervisory role adopted. Alternatively, when there are two observers, one may focus on the above, whilst the other focuses on the creative techniques used and the impact on the supervisory relationship.

Each small group practice session is followed by a peer discussion in which these observations are explored. The process of giving and receiving authentic feedback in an acceptable and useful way relates directly to the developing competencies of the supervisor.

Example of triad practice:
The supervisee brings a dilemma about whether to take on a piece of work that has been offered, but that will require her to travel and may have an impact on her energy and stress levels around the rest of her practice. The supervisor teases out the dilemma and together they agree to a supervisory question: How would taking on this piece of work impact on my self-care and my ability to function well in my current practice, and what would be the implications of saying yes to the offer.

The supervisor offers a small-world sculpture technique to map out the supervisee's current working week, and to look at the change that would be made by taking on the extra piece of work and the journey it would entail.

In viewing the two sculptures in miniature the discussion centres around how attractive it is to take on the new work, which is with a different client group and a well thought of organization. It would look good on the practitioner's CV and could lead to further, interesting work. On the other hand it would entail a late night followed by an early morning next day…on a regular basis.

The supervisor asks the supervisee how much flexibility there might be in other parts of the week to make time for self-care if the supervisee takes on the new piece of work.

The outcome is that the supervisee feels clearer about the cost to her private life of taking on the new piece of work, but acknowledges that she is at a point in her career where she is developing her practice and is inclined to work towards a closure with a less rewarding piece of work she has with a different agency over the next six months rather than let this opportunity go. Knowing that the extra energy cost will be finite leaves her feeling inclined to take the offer of the new piece of work.
The sculpture is de-roled and the session brought to a conclusion.

The observer has been taking brief notes and now feeds back to the supervisor what she observed. The quality of listening is reflected on, and the definition of the supervisory question. The use of small world is seen as a good use of the method, allowing an overview of a complex situation and an easy way to view the impact of changes. The symbolism of self-care as a pair of small brass scales could have been used more, e.g. by speaking from the scales when unbalanced one way, then the other way, and what it would mean to be in balance.

The temptation is for the observer to get involved in the content of the session, rather than the role of the supervisor in response to the situation presented. There may be a place for 'sharing' in the psychodramatic sense, but not at the expense of the training feedback to the supervisor.

5.2 Fishbowl

It is to be expected that there is at first a certain degree of vulnerability around the practice sessions. There is a high level of self-disclosure, both explicitly in terms of content of material presented as supervisee, but also implicitly for the supervisor, as the neophyte supervisor's skills and 'mistakes' are observed live by peers. It is essential that the group participates in creating its own learning contract at the start of the course, designed to create a helpful and respectful learning environment. The right to make mistakes is usually high on the list.

Once the learning environment is experienced as safe, and the Triad work has bedded in we offer a further practice structure to the group: the fishbowl. The task of supervising is so complex, and we demand such a high degree of self-awareness and reflexivity that having the whole group on hand with specific observer tasks can be immensely supportive and provide rich opportunity for learning.

We have a maximum group size of twelve, so in fishbowl there are up to ten 'specialist observers'. They witness the whole session from the perspective of one of the seven eyes, or from the perspective of supervisory roles, supervisory question, or with a view to non-verbal communication, metaphor, and use of creative methods. The whole group discussion that follows reveals the multiplicity of ways a supervision session can be viewed, the roads taken and the roads not taken within the session.

A further variation of the fishbowl is used when practicing group supervision. We may structure this with half the group inside as supervisees and the other half of the group as the reflective fishbowl. More common is to have the whole group as supervisees, apart from the person practicing their group supervision skills. In this case the fishbowl is 'virtual', symbolized by empty chairs, representing the participant observer perspective within each group member.

Processing the work in this case has to be highly structured, differentiating sharing-type feedback from more cognitive level reflections. Chairs may be named as "Something I experienced…", "Something I noticed…", "Something I might have done differently…" Alternatively a number of chairs representing the different eyes/lenses are used and the group encouraged to make statements from these. E.g. "A moment when eye five (parallel process) was in the forefront was..." In each case observations are offered concisely from the relevant chair/role and the process remains dynamic.

Example of supervising an individual in Fishbowl:

Trainee Supervisor: sets the scene 'as if' the observing group were not there, i.e. sets up the space and props in a way suitable for the supervisee to use for a session. The trainee supervisor is responsible for timekeeping within the limit set e.g. 30 or 40 or 50 minutes (on the basis that supervision sessions do not have be devoted to one piece of work, but may address a number of different foci in one session).

Supervisee: presents a real clinical or work related issue of genuine concern.

Trainee Supervisor: teases out a focus or supervisory question, clarifies it and chooses an appropriate action technique if applicable to explore this.

Observer 1: Listens from the perspective of the supervisory question: was it identified, was it a helpful focus for the material the supervisee came with, and did the supervisor serve it in his or her interventions?

Example of post-session feedback: The supervisee came in talking about her concern about one client: a ten year old child with whom she is working on the sudden death of his mother. The supervisor identified the supervisory question as "How directive should I be with this work, bearing in mind that the child will be changing school at the end of the academic year?" This remained a central focus of the supervision session, and seems to have been helpful.

Observer 2: Listens from the perspective of eye 1: i.e. what does the supervisee report about 'what happened' in their work, and to what extent was this the focus of the supervisor's attention.

Example of post-session feedback: I noticed that we never found out how and by whom the client was referred for therapy and how explicitly the time limitations of the work have been acknowledge between them. On the other hand they did give a clear account of what happened in a session where the boy made a story about a boy who is kidnapped and taken into the mountains by bandits.

Observer 3: Listens from the perspective eye 2: i.e. what does the supervisee report/ask about interventions and strategies, and to what extent was this the focus of the supervisor's attention.

Example of post-session feedback: I believe that this eye is central to the supervisory question, ie what to do, how to work with this boy given the time limitations of the frame. By the end of the session she is going away having formulated how she might mention the issue of the loss of his mother in the next session, and she has teased out the possible benefits and problems with taking this strategy.

Observer 4: Listens from the perspective of eye 3: i.e. what does the supervisee report/ ask/ explore about the client's transference towards her, and the nature of the therapeutic alliance, and to what extent was this the focus of the supervisor's attention.

Example of post-session feedback: I think this was a missed opportunity. As a female therapist working with a boy who has lost his mother I would have liked to have heard some discussion about what she represents for him transferentially, and how fear of losing her might make him compliant or cautious in the sessions.

Observer 5: Listens from the perspective of eye 4: i.e. what does the supervisee report/ ask/ explore about her own countertransference towards the client, or how the work impacts on her and to what extent was this the focus of the supervisor's attention.

Example of post-session feedback: There was a moment during the role play technique the supervisor offered the supervisee, when the supervisee's voice tone and her body language conveyed a lot of anxiety. I noticed that the supervisor picked this up and said "You seem very cautious about naming the bereavement." I think this went to the heart of the dilemma for this supervisee – maybe it could have been explored further. I wondered about the supervisee's own experiences of loss.

Observer 6: Listens from the perspective of eye 5: i.e. parallel process between the supervision process and the therapy process and to what extent that is taken as the focus by the supervisor.

Example of post-session feedback: I agree with the previous observer. There was a moment when I wondered if the supervisor was protecting the supervisee from thinking about losses in her own life...and I wondered if this paralleled a tendency to err on the side of caution and even avoidance between supervisee and client as regards this painful material. Maybe it would have been helpful if the supervisor had named this dynamic as it was happening.

Observer 7: Listens from the perspective of eye 6: i.e. the supervisor's own countertransference or response to the material, and to what extent this was used in the session.

Example of post-session feedback: I heard the supervisee say, when initiating the role-play "Maybe it would be helpful to bite the bullet and practice different ways of naming explicitly the work you are doing with this child". I was struck by the metaphor of biting the bullet, which suggested to me that the supervisor might be identifying with the supervisee's reluctance to be explicit about the work.

Observer 8: Listens from the perspective of eye 7: i.e. the wider context, and to what extent this was used in the session.

Example of post-session feedback: The wider context here could be seen as the school setting for the therapy, and change of school that is due soon; also the socio-economic conditions of this family that has lost an earner and carer. And of course the whole issue of bereavement and loss (how many other children in his class have lost a parent?) I think that hearing about the referral process would have helped you and us understand more about how he is managing at school. There was a lot of aggression and sadness in the story the child told...I wonder how much of this he expresses in the classroom?

Observer 9: Looks from the perspective of action techniques used and how helpful or appropriate they were.

Example of post-session feedback: The supervisor chose to use role-play and role reversal to explore the possibility of the supervisee addressing the bereavement more directly. I liked how you coached her and gave her permission to 'be insensitive and get it wrong' in the session with you to free up her blocked spontaneity. I thought she got some useful insights from the role of the client, which seemed to give her permission to take a risk.

Observer 10: Looks from the perspective of metaphor and imagery in the session and to what extent it was used as a focus.

Example of post-session feedback: I was bowled over by the child's story that he created and I was hoping that you might explore the roles and dynamics within that story to shed light on how ready the client is to explore these issues directly. I noticed that the supervisee naturally drew upon the imagery of the bandits and enforced separation from a loved one when in role as the client, and I wonder if the supervisor might usefully highlight that as a good 'way in' to the material.

Following the feedback from the specialist observers the supervisee and supervisor also give their reflections. While this structure would be too exposing early on in the training, it is a valuable learning tool for a group with a good level of trust, and highlights the complexity of the task of supervision, and the many opportunities for different choices. In our experience, once the group has an experience of this structure there is enthusiasm to use it again.

5.3 Specialist presentations and book reviews

Twice during the year of training each participant runs an action-based session for the group. The specialist presentation is an opportunity to share knowledge from a position of expertise and experience. Topics may include specific work contexts, e.g. working with children and families in the context of fostering and adoption, working as a psychodramatist within palliative care, working as an art therapist within a forensic setting. Alternatively, where there is a specific area of expertise in terms of a theoretical base this may be the focus of the presentation, e.g. attachment theory and neuroscience.

The rationale for these presentations is threefold. Firstly, it is considered important as a supervisor to be aware of our own strengths and areas of experience, and to be able to communicate our knowledge and experience base adequately. One of the roles of the supervisor is educative, and this presentation encourages trainees to embrace the role of educator/ informer in a creative way. This is particularly helpful within the context of a training course where new skills are being learned. While there may be a sense of vulnerability and challenge around learning new skills, it is important not to lose sight of those areas where we feel confident.

Secondly, it offers a welcome opportunity for peer learning, including for the tutors. The experience base of each cohort is different and it is a delight to hear about and participate in workshop activities relating to areas we may have little or no experience of ourselves. Towards the end of the presentation group members are invited to reflect in small groups on how what they have learned might inform their expectations if they were to supervise someone in the setting described, or using the theoretical base introduced.

Thirdly, it is an opportunity to practice the creative skills at the heart of the training. The application of action methods to different contexts is not necessarily straightforward. The exercise requires a sensitivity to the context of the group, an ability to role reverse with the peer group and on that basis choose an activity or series of activities that are congruent with the learning outcome desired.

Book reviews are approached in a similar way. Each trainee takes on responsibility for a book review from our list of texts on supervision and related issues. The expectation is that he or she uses a combination of didactic and experiential approaches to help the group engage with core themes and with critiquing the text. This approach embeds the reading and learning in the creative and action-oriented culture of the course. Both the book review and the specialist presentation are backed up by the presenting trainee with a document sent to the whole group.

6. Trainee assessment

The mode of assessment reflects our understanding of the process of becoming a supervisor as a developmental process. The first module is entirely self-assessed. Trainees are encouraged to reflect on their learning over the two weekends, focusing on the following questions:

- What have you learned during this module at a theoretical level? What new skills and techniques have you learned or practised?
- What significant feedback from others have you received during this module and how have you responded to it?
- How do you feel about your process and progress during this module?

By prioritising self-assessment during the first stage we aim to foster realistic and balanced reflections as to achievements and challenges. It is an act of trust and confidence in the trainee, which forms a foundation for the challenges to come. We keep a copy of the trainees' self-assessment document, with an eye for congruence of self-reflection with what we have observed in the group.

The second module is assessed through peer assessment as well as self-assessment. The task of giving peer assessment can be daunting, but corresponds with the need to develop an adequate evaluator role as a supervisor. For therapists, who prioritise being 'non-judgmental' the evaluator role can offer a major challenge as they grow into the role of supervisor. So the training group is used as a practice ground for this role. While the triad and small group practice involve verbal feedback immediately after the practice, peer assessment is about summarising strengths and learning needs succinctly in writing. As practising supervisors we may be asked to give periodic written reports on our supervisees, whether for their training institutions or for employers. So we have a responsibility as a training to help our trainees engage with this task.

The third module, which is chronologically the final taught module, is assessed through self-assessment, peer assessment and tutor assessment. The tutor feedback is presented in the group, transparently. It is brief and is structured in two parts: Strengths and Things to Work on. Again we are looking for congruence between tutor, peer and self-assessment.

There is then an arc to the assessment process in terms of the training weekends. As the trainee group gains confidence and competence the stakes get higher. The final part of the tutor assessment takes place after the end of the taught modules, when a final essay is submitted, linking theory to practice. The essay is an opportunity to articulate how the learning on the course has informed one or more supervisory relationships outside of the training. It is presented with a log of supervision on supervision sessions, signed by the external supervisor on supervision.

The developmental model of assessment reflects an acknowledgment that clinical as well as supervisory practice is about learning and development. Just as we expect supervisors to bear in mind the developmental level of their supervisees and choose their supervisory roles accordingly, so we as a staff team aim to pitch the assessment process to the growing ability of the supervisors to reflect critically on their competencies and limitations without losing the confidence necessary to keep practising and learning.

7. Course evaluation

At the end of each module we ask trainees to complete an evaluation of the course content and delivery in terms of the following criteria: Appropriateness of material; Balance of theory and practical; New learning and ideas; Clarity and knowledge-base of facilitators; Usefulness in work setting; other comments. We aim to model a quality of openness to feedback: we need to know what we are doing right, and if the trainee has leaning needs that are not being met.

8. Creative action techniques

The creative approaches are perhaps the most attractive elements of the training, but they need to be located within a helpful theoretical frame regarding the process of supervision and used with discretion and purpose (Chesner, 2007). As each creative technique is taught we put emphasis on exploring what that technique is good for and when it might be contraindicated. For example role reversal and miniature sculpting have a very different effect in terms of distance and closeness. While role reversal with a client might help a supervisee get under the skin of someone they struggle to understand or empathise with, miniature 'small world' sculpting can be used to focus the lens of attention on the bigger system, and to gain distance from a situation or client experienced as overwhelming. These are some of the techniques taught on the course. They are not comprehensive or fixed, but form a starting point for a number of different approaches:

- Small world, miniature sculpting (see Lahad, 2000): the use of small objects to create an image of a relationship or system being explored.
- Communicube (see Casson, 2007) is an extension of the above, using different levels of the communicube to look at multiple perspectives simultaneously.

• Large scale sculpting: either with chairs or, in group supervision, with group members. This technique can be used as a group or team exploration of a theme or situation, and allows a sociodramatic movement to different perspectives and a collaborative engagement with supervisory issues.
• Role taking and role reversal: this technique from psychodrama is used in the context of understanding the other, whether this be a client, a member of a client's social system, a piece of art produced by a client. It may also be used to understand dynamics with a colleague, a setting or an aspect of self as practitioner.
• Psychodrama supervision vignette: this is an extension of role reversal, in which a supervisory question is explored by an individual within a group setting. It requires familiarity with the psychodrama method on the part of the supervisor and typically incorporates doubling, mirror and multiple role reversals.
• Six Shape Supervision Structure (Chesner, Creative Supervision Across Modality, publication pending): this is a paper and pen and movement based structure inspired by Gersie's therapeutic Six-Piece Story Making (see Gersie, 1997), but adapted specifically to the supervision frame. It is a tool for self-supervision, group or individual supervision.
• Four elements: (Zografou, Creative Supervision Across Modality, publication pending): this is a large group, improvisation-based method developed out of playback theatre and using elements of dance movement, projection, poetry and visual art.
• Mandala: (Jennings, 1999): this is a visual technique particularly suited to reflections at a developmental and holistic level. It lends itself to individual and group settings.

9. Role awareness

The emphasis psychodrama places on role as a core concept is intrinsic to the model of this training. From the outset of the course we encourage the language of role. A recent cohort of trainees created the following extensive list of supervisor and supervisee roles out of an exploratory action session on the first weekend of training. Two chairs were on stage, one designating the supervisor role, the other the supervisee role. Sub-roles and legitimate expectations of the roles were concretised with props, and the following list emerged.

Supervisee Roles (course group 2009-10)		
Prepared contributor	Committed participant	Frank mistake owner
Interested explorer	Courageous questioner	Occasional challenger
Responsible caretaker of client safety	Receptive welcomer	Honest terminator
Engaged reflector	Ethical monitor	Discriminating assimilator of advice
Client-centred risk taker	Generous sharer of knowledge	
Confident prioritiser	Balanced acknowledger of strengths and weaknesses	

Supervisor Roles		
Fully present attender	Brave supervisee	Generous sharer of knowledge
Open-hearted welcome	Transparent report writer	Life-long committed learner
Active listener	Astute seer beyond the surface	Conscientious evaluator of practice
Collaborative reframe	Sensitive human being	Respectful power sharer
Sensitive boundary holder	Creative reassurer	Grounded hope-holder
Congruent respecter of limitations	Non-collusive investigator	Sturdy container of hopelessness
Skilful locksmith	Versatile architect/ relationship builder	Gracious inspirer
Fair but passionate advocate	Welcoming participant	Responsible self-carer
Playful accompanist	Ethical monitor	Reliable time-keeper
Appropriate knowledge holder	Occasional suggester of reading/therapy/training	Consistent reviewer
Intrepid intervener	Appropriate feedback giver	Honest Terminator

10. Conclusion

Since starting to write this paper I have been reading L. Dianne Borders' chapter on Principles of Best Practice for Clinical Supervisor Training Programs in *State of the Art in Clinical Supervision* (Culbreth & Brown, 2010). Her chapter is written from an American perspective but her five core principles for best practice in training supervisors are useful for a European supervisor training courses as well. They are: The course should…

• "…address all the core content areas identified in professional standards and the literature" which she summarizes to include roles and functions of supervisors, models of supervision and counselor development, methods, techniques and approaches, relationship dynamics, cultural issues, group approaches to supervision, ethical and legal issues, feedback and assessment methods, administrative skills and research on the field" (p. 128).
• "…include both didactic instruction and supervised practice, concurrently and/or sequentially. Experiential activities should involve direct observation of supervision practice with feedback" (p. 130).
• "…reflect a developmental approach in their content and sequencing" (p. 136).
• "…include instruction in a wide range of supervision methods, techniques, and approaches, with an emphasis on the intentional and flexible use of these approaches" (p. 140).
• "…include instruction in basic principles of learning theory" (p. 142).

Looking over these principles I would concur that they are crucial to creating a balanced training. Our Creative Approaches to Supervision training clearly offers a number of specialised techniques relating to action methods in supervision (principle 4), but the other principles are essential if these are to be used with appropriate spontaneity.

References

Borders, L. D. (2010). Principles of Best Practice for Clinical Supervisor Training Programs. In J. Culbreth & L. Brown, *State of the Art in Clinical Supervision*. New York: Routledge.

Casson, J. (2007). Psychodrama in Miniature. In C. Baim, J. Burmeister and M. Maciel (Eds.), *Psychodrama Advances in Theory and Practice*. Hove: Routledge.

Chesner, A. (1999). Historical Issues and Supervisory Perspectives. In E. Tselikas Portman, *Supervision and Dramatherapy*. London: Jessica Kingsley Publishers.

Chesner, A. (2007). A Passion for Action and Non-action. In R. Shohet, *Passionate Supervision*. London: Jessica Kingsley Publishers.

Chesner, A., & Zografou, L. (publication pending). *Creative Supervision Across Modality*. London: Jessica Kingsley Publishers.

Gersie, A. (1997). *Reflections on Therapeutic Storymaking, the Use of Stories in Groups*. London: Jessica Kingsley Publishers.

Hawkins, P., & Shohet, R. (2000). *Supervision in the Helping Professions*. Buckingham: OUP.

Jennings, S. (1999). Theatre-based Supervision, A supervisory model for multidisciplinary supervisees. In E. Tselikas-Portman, *Supervision and Dramatherapy*, London: Jessica Kingsley Publishers.

Lahad, M. (2000). *Creative Supervision*. London: Jessica Kingsley Publishers.

Williams, A. (1995). *Visual and Active Supervision*. London: W. W. Norton & Company.

Supervision and practitioner research of trainees

Pierre Fontaine

Having been responsible (1999-2010) for a 'course on research' in the common final year of psychodrama training of four Dutch and Belgian training institutions I would like to give some reflections on my work to integrate research into training. Therefore I want to describe some attempts to motivate trainees to do research from the beginning of their practice.

In this chapter I will firstly describe the position of psychodrama in the field of health. Secondly a model of doing practitioner research for trainees and a bottom-up research is presented. Thirdly it will be shown that supervision can be used to bring students' interest and some basic knowledge about research into practice. Finally the results and some practical examples in practitioner research with trainees will be presented and discussed.

1. Psychodrama and research in the field of health

Research in psychodrama is more a human or – as Moreno (1953, p. 64) put it – a social science where research objects are at the same time research actors.
The position of psychodrama psychotherapy and research in relation to medical approaches in the field of health can be illustrated by a model of Fourez (1992). The two axes of coordinates create a map integrating the type of therapeutic intervention and the level of intervention. The vertical axis reaches from the 'atomic' to the 'cosmic' level and the horizontal one from 'eradicating diseases' on the far left position going to the right side that expresses the 'developing of strengths'. The horizontal axis also includes the attitude of the patient towards a psychotherapeutic treatment from a passive position on the left and an active position on the right.
Psychodrama can be located in this scheme between the 'individual person' and the 'society' strengthening and exploring future action. Considering the main theoretical and practical approaches psychodrama and medical orientations seem to be different and complementary. Leading medical research is at the 'molecular' and 'disease' level and the interest of scientific medicine goes in a south-west direction. On the other side the action of psychodrama is mostly located in the opposite north-east quarter.

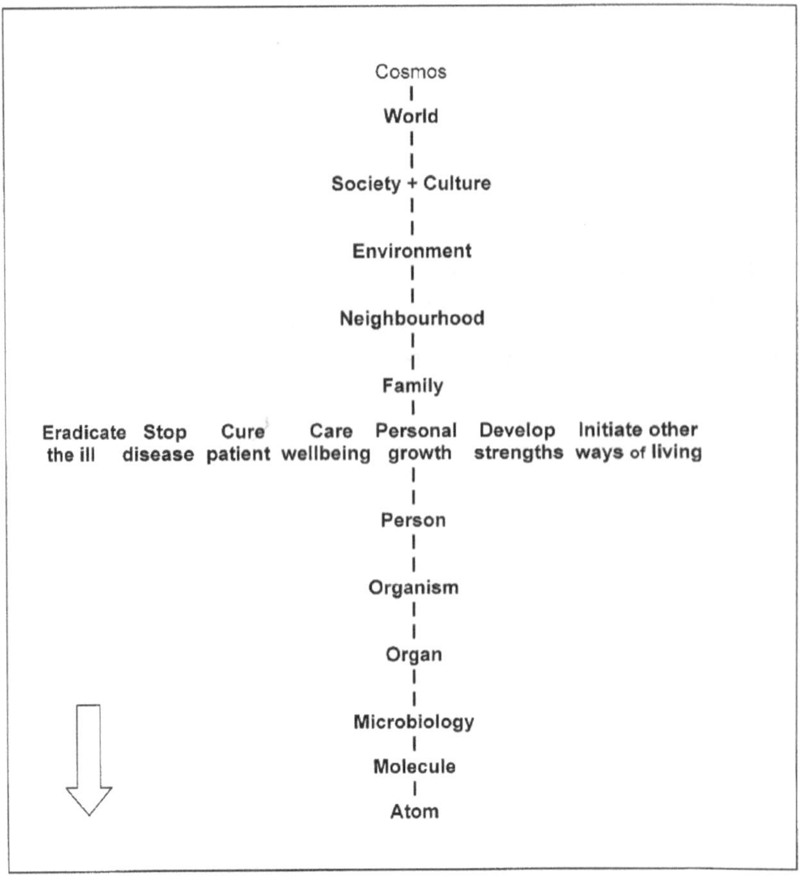

Figure 1: Psychodrama in the field of Health. Vertical axis: fields of medical activity; horizontal axis: ways of intervention. Translated and adapted map of the concept of health according to Lambourne in G. Fourez (1992, p. 92).

2. Supervision as practitioner research

The term research in this context is used in a rather broad sense of its meaning. In order to give students an idea that research is basically always a search for better understanding I began by listening to their ideas. Students should not be overwhelmed by theory and philosophy of science, when their interest is just starting to develop. For this reason practitioner research in this project is seen as an action research of trainees on their own psychodrama practice.

In order to get more practitioners involved in research, trainers need to ensure, that research becomes part of the training from the beginning. As Schön (1983) said we need 'reflection-on-action' and more qualitative research. Regarding counselling research McLeod (2003, p. 192-193) put an emphasis on the following aspects:

- greater awareness of the relationship between research and practice,
- permission to be reflective,
- openness to new methods of inquiry,
- research oriented to discovery rather than verification.

When McLeod (2003, pp.192-193) writes using Schön's (1983) well known metaphor that research needs 'to be grounded in the messy world of everyday practice' and of 'practice based evidence', his ideas meet our understanding of bottom up research. Reflection of own practice ('inner' view) and getting a view from 'outside' can lead to a better understanding. This is part of good supervision and of good research as well. The students learn that the external position gives them a new perspective and that an external observer or group can be helpful.

In my concept of how to introduce research in training we followed a bottom up approach, which can be described in three circles:

- The centre, the heart can be seen as *our own practice*. We question and reflect on our practice with ourselves and our team of involved collaborators, and we can open new horizons.
- Around this centre, there is a wider *circle of colleagues* (who we can meet in professional meetings and by reading their publications) with whom we can share our research and confront our ideas and see if some of them can be generalized.
- Finally, there is the circle of *our society*, where managerial decisions must be taken about training in the profession, introducing new courses, nomination of professors and reimbursement of practices in social security etc. In the society there is also a need to popularise this knowledge for the consumers.

2.1 Context and process of practitioner research

The context in which the students and I worked was that of a final year of training in psychodrama as a 'learning circle' for trainees from various training centres in the Netherlands and Belgium. These courses were born during the meetings of ESCOPE (Education Study and Coordination of Psychodrama Education) where psychodrama trainers met to create a network in order to develop psychodrama training. These meetings led in 1995 to the founding of the Federation of European Psychodrama Training Organisations (FEPTO).

Around 1999 I was called to give a course on research. The first year, I gave articles to the students concerning various forms of research relevant to psychodrama as well as examples of simple research carried out in practice. On these attempts my success in increasing their motivation to do research was rather poor.

In the next cycle, two or three years later, I started the course with reading an e-mail (May 22nd, 2002) sent by David Kipper complaining about the lack of research and clinical studies in psychodrama. After reading I asked the students to show in a spectrogram, where they saw themselves with regard to their attitude towards research. The students realized that they have questions about their own practice. They decided to observe more accurately what they were doing until the next meeting in order to get their questions clearer. They learned that these questions were the basis for beginning research on their own practice. Everyone then started an inquiry and action on a topic from their own practice and wrote a paper on it at the end.

In our programme I had officially only a 4 hours course or workshop to introduce the subject research. We worked together in the following way:

• *Preparation:* Before class, the trainees were given a few introductory texts (Burmeister, 1999; Fontaine, 1999, et al.) regarding research in psychodrama practice and they were asked to reflect on their practice and to choose and propose a topic for their research.

• *Discussion:* First they discussed their topic with their colleagues and they were asked to send in their papers. After that we had a 3 hours meeting, where they discussed the text with me.

• *Action:* Beside the discussion we worked actively to make them aware of their fears and their motivation for doing research. Using the sociometric tool of a spectrogram they were asked to show their position regarding the questions: "What attracts me?" and "What frightens me?" (e.g. "I'm afraid to distort my contacts with my clients", "It takes too much time", see the box 'Spectrogram')

• *Presentation of the project*: At the end the students briefly presented the section of their practice and the research subject they wanted to work on. The group and the facilitator gave feedback on it. Within a month they had to send a paper of 2-3 pages about their research project with a request for acceptance.

• *Supervision:* During their work they obtained supervision by phone about their research. Beside that the trainees shared regularly in their learning groups with their colleagues and received one hour more with me as their supervisor.
• *Presentation of results*: Finally they sent a paper of about a dozen pages. In half of the cases, it was returned with some opinions, questions and suggestions. The final version was read and assessed by three different teachers.

Spectrogram of hindering and attracting aspects:
What encourages or hinders research? The trainees' feelings can be expressed through dramatic play: two chairs, spectrogram, satellite, meeting of psychodrama and research... Here we give some expressions from a spectrogram:

Hindering aspects:
- We are too small, only great psychodramatists are doing research. We are not academics.
- Research will cut us off from our client; will be an intrusion in our relationship.
- It will ask for an investment of time. At first it might be fun to work on a project, but when it progresses and we have to comply with the rules, it will become boring.
- In the role of psychodrama:"I am a full girl, free and spontaneous. Don't put me-behind grids for analysis."

Attracting aspects:
- Research would allow us to find our professional identity better, we affirm ourselves to others.
- Research in our practice is possible. Qualitative research makes sense and is increasingly accepted.
- Our client picks up the truth in psychodrama and we, the psychodramatists can find our truth in research.

2.2 Research projects of trainees – some examples

The following examples are reflections of trainees on their practical work. They show two types of research in practice: firstly a questioning of and reflection on own practice of psychodrama, and secondly one type of reflected innovation and exploration of new fields or forms of practice.

Marianne Mostert (2008) wanted to know why she is so often prolonging her sessions with the clients. By questioning herself why she could not finish in time she found out that she wanted her clients to appreciate her work. She worked longer to be successful in reaching this aim. In her work she described the way she changed the preparation for the sessions and her personal warm up. By reflection she found a procedure that changed her attitude.

Cécile Osse (2008) an experienced consultant in using sociodrama as a helpful tool to work with merging companies wanted to know if it would be also helpful to use sociodrama in the integration process of immigrants. Sociodrama workshops with immigrants were carried out and two observers described the sessions in order to evaluate the process.

Carlo Lesage (2009) an art director who was working for years in helping amateur actors to set up a play each year wanted to know if psychodrama could be helpful in inspiring actors to create a good show. In his paper he described the process when he encouraged the actors not to start with a script as they were used to but to play pieces of their own life. At the end he discovered that by doing so he changed his aim. Not the most entertaining play was presented but the one that has the most healing effect on the actor.

Marjolijn Van Burik (2002) compared two methods of sharing experiences and evaluation in a personal training group for a company. At the end she asked the participants about the effects of the new approach.

In the first method the participants were asked one by one to describe the experiences they had in a previous training group. In the second method they formed pairs and talked about their experiences to their partner. Later in the group again each of them took up the role of the partner and told the group the experiences as if they had been their own. The feedback given after the exercise showed that the participants felt better with the new approach.

Moïra Verhofstadt (2001) practiced individual psychotherapy with people convicted of sexual abuse. She started to work with certain clients using psychodrama in individual sessions without knowing why she had chosen them and not others. She wanted to know more about what aspects influenced her decisions. She used lists of factors that were related to the therapist, the client, their relationship or circumstances and rated the influence of this aspect on her decision.

The goals for this project were comparable with my goals as a supervisor. I wanted to facilitate the first steps in practice by training the practitioners and helping them with their practice. In order to reach these aims, the instrumental goals for me were (a) to improve their practice by reflection on it, and (b) to find, generate and choose more adequate alternatives.

Some of the students remained with their work in the so called 'first circle' of research as a reflection on their own work. Some went beyond their own practice and reached what I called the 'second circle'. One study was published in an international psychotherapy journal. It reached a large circle of colleagues. Cecile Osse's way of working with sociodrama was demonstrated at a meeting about problems of integration of migrants with mental health professionals and politicians. The psychodrama-theatre was presented to different groups of psychodramatists but also many times in public. These projects partially reached the society and what we called the 'third circle' of research.

The trainees were motivated and invested much time and creativity in their own project. They developed their professional identity and their flexibility and creativity in different subjects. These examples of research illustrate that an open approach to the topic of research allows student with a non-scientific background to get in touch with the role of the researcher.

3. Conclusions

The openness to new methods of inquiry allows us to be creative and not to be caught in models of research which are already well developed. The experiential approach seems to be a motivating aspect for the trainees and it comes close to their identity as creative psychodramatists.

Considering the development from the first attempts to the last process it can be said that trainees in general like to reflect on their own work and to learn by investigating their practice. Action methods turned out to be helpful in increasing their motivation to reflect on their practice. It can be a first step and initial process that encourage trainees to be researchers on a basic level.

Supervision of practice during training is a good moment to let students experience the role of a researcher. It starts always with a question and leads from simple description of practice to more complex methods of systematic inquiry.

References

Burmeister, J. (1999). Research and psychodrama training. In P. Fontaine (Ed.), *Psychodrama training. A European view* (pp. 275 – 287). Leuven: FEPTO Publications.

Fontaine, P. J. (1999). A spirit of research in psychodrama training. In P. Fontaine (Ed.), *Psychodrama training. A European view* (pp. 289 – 299). Leuven: FEPTO Publications.

Fourez, G. (1992). *La construction des sciences. Les logiques des innovations scientifiques*. Bruxelles: De Boeck.

Kipper, D. (2002). Personal e-mail received 22 May 2002.

Lesage, C. (2009). *Zelfonderzoek binnen mijn werk als theatermaker.* [*Self-research on my job as theatre producer*] Onderzoek Learning Circle CP 5 of NBES.

McLeod, J. (2003). *Doing Counseling Research* (2nd ed.). London: Sage.

Moreno, J. L. (1953). *Who shall survive. Foundation of Sociometry, Group Psychotherapy and Sociodrama.* Beacon (N.Y.): Beacon House.

Mostert, M. (2008). *Tijdwinst door warming-up.* [*Time saving through warming up*] Research Learning Circle CP 5 of NBES.

Osse, C. (2008). *Sociodrama een succesvolle benadering voor het oplossen van sociale conflicten !? [Sociodrama, a successful approach to solve social conflict]* Onderzoek Learning Circle CP 5 of NBES.

Schön, D. A. (1983). *The Reflective Practitioner. How professionals think in action.* London: Temple Smith.

Smith, M. K. (2011). Donald Schön: learning, reflection and change. *The encyclopedia of informal education.* Retrieved from http://www.infed.org/thinkers/et-schon.htm. 6 Dec 2011.

Van Burik, M. (2002) (in Dutch). [*Evaluation by role reversal and doubling in pairs*] Onderzoek Learning Circle CP 4 of NBES.

Verhofstadt, M. (2001). *Wanneer is het aangewezen tot psychodramatechnieken over te gaan bij een bepaalde client en wanneer niet ?* [*Indications and contra-indications for psychodrama techniques for a precise client.*] Onderzoek Learning Circle CP 3 of NBES.

Promoting research-practice: supervision as an ideal moment to develop the role of a psychodramatist-researcher

Gabriela Moita & António Roma-Torres

Research in psychodrama is commonly referred to as a necessity. Psychodrama, like the vast majority of psychotherapies, is far from achieving a desirable level of empirical validation of its effectiveness and there is still a lot of research to be done on the therapeutic process. In order to foster research, this paper proposes the integration of research skills in the training programme of the psychodramatist.

In this article we present an experience of teaching research methodologies and monitoring of project development during supervision in psychodrama. It is argued that the phase of supervision is the ideal moment for such an acquisition as the results of research are themselves a response to the doubts of the new psychodramatist. It is also argued that this stage of role taking allows the psychodramatist to develop the role of psychodramatist-researcher, which will facilitate continuity in the conduct of future research.

The research model proposed is the Hermeneutic single-case efficacy design (HSCED), according to Robert Elliott (2002), a mixed methodology that combines qualitative and quantitative approaches, permitting the assessment of both outcome and process.

1. Introduction

The need for research in psychotherapeutic practice is emphasised by both epistemologists and politicians. For the former it is the only way to guarantee the scientific status of psychotherapy, while politicians seek fundamentally to ensure its effectiveness in order to justify costs, requiring empirical validation. Psychodrama is not and must not become an exception to this requirement. Wieser (2007), in his meta-analysis studies, clearly states that "The scientific status of psychodrama psychotherapy has not yet been recognized by the scientific community at large, since studies done in this field seem to have failed to attain mainstream standards" (p. 271). In this sense we suggest that research skills should be incorporated as a skill to be acquired in psychodrama training programmes for psychotherapists. To this end, we present an experience of teaching research skills developed during supervision.

We believe that carrying out researched psychodramatic practice offers a way of overcoming the scarcity of currently available research. The phase of supervision is typically a moment where trainees pay special attention to the effect of their intervention, and therefore it seems the moment to promote the development of knowledge and research skills, stimulating the search for response to this effect through empirical validation. Research-practice meets the needs of the therapist, of his or her field of knowledge, as well as the demands of governments and, most importantly, it responds to the patient's needs.

A brief historical note sets out some reflections on research in psychotherapy. Subsequently, we present the work with a supervision group. We then share the training programme which was given to students during a workshop organized specifically for this purpose. The methodological framework is succinctly presented as a starting point as well as the proposed instruments. As stated at the end of the presentation, it should be stressed that this plan is also being developed by a workgroup of FEPTO's Research Committee (TRAIN – Towards Research Applied in an International Network of Trainees) and that it is on the way to be implemented in the psychodrama training curriculum in Austria and Turkey[1] (Krall, Doganer, & Fuerst, 2009, p. 6-7).

2. General aspects on psychotherapy research

It is a well-known fact that that a huge gap exists between psychotherapy and research on psychotherapy. The reasons for this gap are the result of various underlying factors, such as:

- the lack of research skills, associated with the idea that research is very difficult and therefore inaccessible,
- the time it takes to do research,
- the idea of research as an isolated field of knowledge, distant from the reality of intervention and with little implication for intervention.

Some of the resistance towards research also focuses on the discussion about methodologies. Some methodologies are underestimated, others are overestimated, be they of a quantitative sort, fitting a positivist paradigm, linked to

[1] See the reports written by Hannes Krall, Inci Doganer and Jutta Fürst, members of this FEPTO RC workgroup at FEPTO Website http://www.fepto.eu/storage/files/documents/TRAIN.pdf

metanarratives, be they of a more qualitative kind, grounded in a post-modern paradigm.

The lack of knowledge of research skills: If we raise questions about the place and time where the research process is taught, we find many shortcomings. In the academic world, in an initial training phase, by virtue of the very generality of teaching programmes, the concepts of research are elementary and rarely linked to a particular object of study. Postgraduate courses are dedicated mostly to know-how. We therefore conclude that learning of the research process is exclusive to academic programmes of masters and doctoral syllabuses, by their nature designed to teach applied learning research. On the other hand, teaching outside the university, provided by associations and scientific societies, as is the case with many postgraduate courses and, in particular, the generality of training in psychodrama, usually does not take into account the teaching and learning of these skills.

The time that research takes: The time an investigation takes depends on the objectives of research, the research design chosen, the instruments selected to collect information, both with regard to its application and to scoring and data analysis. For quantitative data, it is possible to make a rigorous research design using statistically validated instruments, for simple and quick administration. For designs of a more qualitative nature, there are various devices that can be created for data collection. In both cases, a growing number of computer programs facilitate data processing and analysis.

The less immediate the benefits produced by any given piece of research, the less value will be attached to the time dedicated to it. However, many instruments or data collection procedures can be readily used as helpful information in psychotherapeutic practice.

Research as an isolated field of knowledge, far from the reality of the intervention and with little implication for intervention: Some data collection tools employed for research objectives are at the same time useful in guiding psychotherapeutic practice, and can be an excellent resource for the psychodramatists' choice of warm-up, the selection of a protagonist, the appropriate suggestion of enactment. Many instruments used for research purposes or evaluation of efficiency can be employed simultaneously, or even simply, as tools for therapeutic use. In this perspective, the time spent administering them can be regained in the therapeutic process itself.

Célia Sales et al. (2007) attempted to understand how therapists who employ research tools in their routine perceive this practice. They discussed in particular the perception of the use of two instruments that provide qualitative data and which will be referred to later, the PQ – The Simplified Personal Questionnaire (Elliott, Mack, & Shapiro, 1999) and HAT – Helpful Aspects of Therapy (Llewelyn, 1988). Besides a list of the advantages and disadvantages perceived by therapists, and also of advantages and disadvantages for the client perceived by therapists, we wish to stress here the fact that most therapists participating in the study state that they would like to integrate information obtained from PQ and/or HAT in their clinical work (PQ - 92%; HAT - 91%). It is also worth noting that the therapists who took part in the study felt that the use-value of these instruments outweighed any possible negative interference.

In the history of psychotherapy research three major moments can be identified according to the development and evaluation of different methodological paradigms (Moreira, Gonçalves, & Butler, 2005). The first phase, prior to 1952, is a period where psychology is still striving to assert itself as a science and is therefore defined by an overriding need for the application of experimental methods to answer the question "Is Psychotherapy a scientific activity?" Accordingly, the subject of investigation is the result, focusing on the therapist's behaviour in an analytical perspective, since it is argued that changing the client is a dependent variable of the therapist's behaviour. The instruments used for data collection focused therefore mainly on the action of the therapist. Researchers resorted to quantitative designs to ensure accuracy and operational objectivity (principle of variable control).

A second phase, between 1952 and 1969, followed the publication of Eysenck's article "The effects of Psychotherapy: an evaluation" (1952). In this article, Eysenck calls into question the validity of psychotherapy. He suggested that psychotherapy had no positive effects, and, conversely, could imply negative effects and certain damage to the client. This disqualification of psychotherapy led to a research boom as a response. On the other hand, in the post-war era the United States government required a guarantee that the therapeutic process was effective in order to reimburse the costs of ex-soldiers' psychological counselling. This interest in the results materialized in the funding of research on the effectiveness of psychotherapy. A shift of perspective occured then from the analysis of the therapist's behaviour to the evaluation of the client's behaviour. A growing concern and an increasing demand on the research process itself becomes clear, including: the evaluation of the research process, the refinement of research methodologies, the standardization and operationalization of models, the development of manuals, the construction of standardized measures to assess outcomes, the use of statistical techniques to evaluate the clinical significance of the effects of psychotherapy, randomization, among others (Moreira, Gonçalves, & Beutler, 2005).

The third phase of research in psychotherapy is marked by the constitution of the Society for Psychotherapy Research (SPR) in 1969, a multidisciplinary organization that seeks to define criteria for investigating the effectiveness of psychotherapy. In this third phase a growing importance is ascribed to the object of investigation itself. In addition to the assessment results, the current concern is to evaluate processes and understand the interactions between clients and therapists; the perspective of analysis is now defined by internal variables that mediate the therapeutic process, as well as mechanisms of change. There is a proliferation of mixed methodologies, both quantitative and qualitative. This stage also witnesses the development of meta-analysis. Meta-analysis aggregates the data and results of a set of studies that compare the effects of psychotherapy and control groups, analysing them through statistical procedures. As a result, it has been stressed that the subjects of the control groups showed an improvement rate of 30% while the subjects who underwent a therapeutic process improved by 70%.

This division into three historical moments corresponds to the development of major areas of research in psychotherapy that are synthesized by Elliott (2002), induced by the issues that motivate intervention; Elliott notes, respectively, three areas of research:

- Psychotherapy outcome research,
- Psychotherapy efficacy and effectiveness research and
- Psychotherapy change process research.

These territories are defined by the issues that prompt research: Has this client (or group of clients) actually changed? (*Psychotherapy outcome research*, e.g., Strupp, Horowitz, & Lambert, 1997), is psychotherapy generally responsible for change? (*Psychotherapy efficacy and effectiveness research*; e.g., Haaga & Stiles, 2000), and what specific factors (within therapy or outside it) are responsible for change? (*Psychotherapy change process research*, e.g., Greenberg, 1986) (as cited in Elliott, 2002, p. 1).

This brief note on the history of psychotherapy research allows us to observe that we are at a point where there is a greater openness to the demands of research, the most important being study through a consistent design which leads to conclusions about the effectiveness and/or understanding of aspects of the therapeutic process. Each one of these approaches allows an advance in knowledge within the field of psychotherapy. The concern should not be to find the right method to investigate, but rather the selection of an appropriate method to answer the question. Depending on the research issues, the best option can be either a qualitative or a quantitative methodology, or even a combination of both. Our era is one of rupture with quantitative methodologies which have so far, for historical and social reasons, come to dominate the field.

Moreno was not oblivious to this need, having dedicated himself to research since he began working in the United States in the 1930s, starting in Sing-Sing prison and continuing later in Hudson. He resolved his "measurement crisis" (Bustos, 1979, p.13) with the creation of sociometry.[2] He defined its object of research: human relationships, more specifically the links established between the elements of a group. He developed its own measurement instruments: the sociometric test, the test of spontaneity. Unfortunately, he did not have many followers in this role as psychodramatist-researcher. The development of methods to be offered to the psychodramatic field of research, both for psychodrama itself as for other research models, is an area still to be fostered.

If we want psychodrama to be validated as a model of psychotherapeutic intervention with epistemological openness, it is important to contribute to the investigation of our practice. Studies aiming to answer questions that help us better understand the model itself are relevant and necessary so we can provide new knowledge to the field of psychodrama, according to research questions, using all types of design (case studies, studies on the efficiency of techniques, randomized studies using control groups, studies of process). However, to boost research in psychodrama it is necessary to endow psychodramatists with the necessary skills.

3. The integration of research skills in the supervision stage of psychodrama training

We consider the period of supervision the ideal moment during the training process to encourage and stimulate our students with the research of their therapeutic practice. The intervention itself benefits from the findings as they are being obtained through research, and both therapist and patient obtain a more objective feedback of the process.

During the time of supervision, in this early stage of role taking, trainees need more feedback, and they seek it. To facilitate their getting acquainted with research methodologies at this stage, where the need is felt, means to offer them an evaluation of the results of their own intervention based on empirical evidence. It helps them, Valerie Brito (2006) states, to "investigate for themselves the principles and concepts they want to apply [in] their work" (p. 16). We thus give support to the role of psychodramatist by promoting autonomy and

[2] Only later will sociometry be developed by Moreno as the core of a theory. "Sociometry was born to serve its purpose of research and measurement of interpersonal relations. Only later does Moreno propose sociometry as a theoretical landmark of psychodrama" (Bustos, 1979, p. 15).

confidence, and facilitate the acquisition of skills for the development of the role of researcher, allowing us to create the role of psychodramatist-researcher.

Psychodrama can provide the research field with new methods of data collection, which will not only develop research in psychodrama, but also facilitate the investigation of other therapeutic models as well as benefiting research areas besides psychotherapy. An example of this is sociometry. However, only from the role of the researcher may this research field be brought to bear. It is important to promote a warm-up for the emergence of this role.

The awareness of the importance of research and the interest in doing research in psychodrama was expressed by some of the trainees in the first sessions of supervision with one of our supervisory groups. The authors' attention to this need is associated to the fact that they have promoted the discussion on this issue in the FEPTO Research Committee. We were both, although at different times, actively engaged in this Committee, and assume this motivation as a crucial one, at a moment we consider to be perfect for psychodrama trainees to promote engagement with their own research questions. It is this work developed with trainees we would like to share here.

The very concept of research, as well as notions of instrument, methodology, research plan, the subject of research, outcome studies, and studies of process were raised and explored on stage, with the use of statues, soliloquies, doubles, and small vignettes.

Trainees were offered a workshop, held by Célia Sales, an expert in family therapy research, on theoretical research in psychodrama, with the presentation of a historical review of research in psychotherapy; futhermore, we acquainted trainees with the instruments of a methodological approach we proposed. In addition to providing a research design, we facilitated the contact with the instruments through role play to promote experiencing and solving questions.

The proposed research methodology was the hermeneutic single-case efficacy design (HSCED) proposed by Robert Elliott (2002), a design that combines tools for quantitative data collection and analysis with tools for qualitative data collection and analysis, which allows users to carry out comparative studies while at the same time paying attention to the specificity that each situation deserves. This is a design which combines an assessment of results along with the evaluation of process. It is worth mentioning that these instruments can be used both as research tools and for intervention purposes.

The instruments of outcome assessment we proposed were the CORE-OM (Evans et al., 2000), a standardized instrument, and the PQ (Personal Questionnaire; Elliott, Mack, & Shapiro, 1999) a questionnaire that, as the name implies, is built according to each patient's issues and together with the patient.

The instruments of process measurement we adopted were a questionnaire containing open questions that focused on the aspects considered useful during the session (HAT – Helpful Aspects of Therapy; Llewelyn, 1988) and an interview assessing change (Client Change Interview; Elliott, 1996). Our proposal is to add to this design some tools that would assess psychodramatic factors. As an example, we intend to administrate the Spontaneity Assessment Inventory (SAI-R – Spontaneity Assessment Inventory; Kipper & Hundal, 2005) along with some measure of group cohesion, yet to be decided.

This proposal was presented to trainees as a starting point. Trainees were asked to bring instruments they themselves had already had contact with and that they considered to be helpful. We started to build a specifically created instrument data base to assess the effectiveness of psychodrama. Adaptations were made to this basic research design to groups and according to the supervisory objectives drawn by us. The student group also conducted a survey on prior investigation carried out in Portugal in the field of psychodrama, thus initiating the creation of a research database.

At a time when doubts about the effectiveness of the intervention are more pressing than ever, this proposal is allowing us through practical research, to answer more questions and to achieve a greater degree of certainty, both through experience (role taking) and empirical evidence. On the other hand, to assume the role of researcher, even if it be for a short period, allows those involved to "demystify means and ways of research and especially to be more critical of the results and their possible applications" (Brito, 2006, p. 19). We believe this is a way to contribute to the adoption by the psychodramatist of "a scientific attitude of questioning in everyday life [in] their professional actions" (Brito, 2006, p. 20).

To conclude, we would like to reiterate that this same model is being developed by a working-group of FEPTO's Research Committee. In the near future, it will be a model to be offered to different institutes, societies and associations that integrate FEPTO (Federation of European Psychodrama Training Organizations).

We would like, by sharing this line of work, to contribute to the promotion of research in psychodrama. Making psychodrama trainees acquainted with research skills is, we believe, a way to encourage and promote the development of the role of psychodramatist-researcher. Everything leads us to believe that the earlier psychodramatic practice is subjected to empirical evaluation, the more easily research-practice will become a working method in the future.

The moment of greatest readiness to receive any information is precisely the moment where that information allows trainees to meet their needs. At the supervision stage students are, par excellence, at a time of doubts about the effectiveness of their work. Any information that allows them to answer this question will, we believe, constitute a positive gain and will be sought with an ideal level of motivation.

One of the limitations of the experiment described here is that the methodologies and research methods proposed were not in themselves psychodramatic. However, if psychodrama can offer research methods for the assessment of other practices, it should also be able to accept being assessed by extrinsic methods, thus avoiding epistemological isolation. At the same time, by fostering the development of the role of psychodramatist-researcher we are creating the conditions that enable psychodramatists to focus on the development of psychodramatic methodologies.

We believe that by promoting research skills during their training, the student of psychodrama is encouraged to investigate the practice. Thus, trainees obtain personal advantage: by answering their questions, they contribute to the investigation of psychodrama, providing empirical validation and helping to reduce the current gap between psychotherapy research and psychotherapy.

References

Brito, V. (2006). Um convite à pesquisa: epistemologia qualitativa e psicodrama. In A.M. Monteiro, D. Merengué & V. Brito (Eds.), *Pesquisa qualitativa e psicodrama* (pp.14-56). São Paulo: Editora Ágora.

Bustos, D. (1979). *O teste sociométrico. Fundamentos, técnica e aplicações.* São Paulo: Editora brasiliense.

CORE System Group, (1998). *CORE System (Information Management) Handbook.* Leeds: CORE System Group.

CORE System Group, (1999). *The CORE System User Manual.* Leeds: CORE System Group.

Elliott, R. (1996). *Client interview schedule change.* Unpublished research instruments, University of Toledo.

Elliott, R., Mack, C., & Shapiro, D. (1999). Simplified Personnal Questionaire procedure. Retrieved from http://www.experientialresearches.org/instruments.html [Oct. 2, 2010].

Elliott, R. (2002). Hermeneutic single case efficacy design. *Psychotherapy Research 12*, 1-20.

Evans, C., Mellor-Clark, J. Margison, F. Barkham, M., Audin, K., Connell, J., & McGrath, G. (2000). CORE: Clinical outcomes in routine evaluation. *Journal of Mental Health, 9.3*, 247-255.

Goncalves, O. F. (March 2007). O regresso à ciência básica. Editorial do *Boletim Informativo de Psicologia*, Nº *3* (2), Retrieved from http://www.dicas.sas.uminho.pt/uploads/BIPsi%20%20of%202007.pdf20Fevereiro% [April 15, 2007].

Kipper, D. A. & Hundal, J. (2005). The Spontaneity Assessment Inventory (SAI): The relationship Between nonspontaneity and spontaneity. *Journal of Group Psychotherapy, Psychodrama and Sociometry, 58*, 119-129.

Kipper, D. A., & Shemer, H. (2006). The Revised Spontaneity Assessment Inventory (SAI), Spontaneity, Well-Being and Stress. *Journal of Group Psychotherapy, Psychodrama and Sociometry*, *59*, 127-136.

Krall, H., Fuerst, J., & Doganer, I. (2009). *TRAIN - Towards Research in an International Network of Trainees*. In Minutes of the Catania FEPTO RC Meeting October 16-18, 2009, Retrieved from http://www.fepto.eu/storage/files/documents/TRAIN.pdf

Llewelyn, S. (1988). Psychological therapy as viewed by clients and therapists. *British Journal of Clinical Psychology*, *27*, 223-238.

Monteiro, A.M., Merengué, D. & Brito, V. (2006). *Pesquisa Qualitativa e Psicodrama*. São Paulo: Editora Ágora.

Moreira, P., Gonçalves, O. & Beutler, L. E. (2005). *Métodos de Selecção de Tratamento*. Porto: Porto Editora.

Sales, C. et al. (2007). Psychotherapists Openness to Routine Naturalistic Idiographic Research? *Health and Learning Disabilities Research and Practice*, 145 -161.

Wieser, M. (2007). Studies on treatment effects of psychodrama psychotherapy. In C. Baim, J. Burmeister & M. Maciel (Eds.), *Psychodrama: Advances in theory and practice* (pp. 271-292). New York, NY US: Routledge/Taylor & Francis Group.

A Greek model of supervision. A visional encounter

Ioannis K. Tsegos & Natassa Karapostoli

The 'Greek model of supervision' – also known as 'Group analytic supervision' – is a conception introduced and applied at the Institute of Group Analysis, Athens (IGAA) since 1983 (Tsegos, 1984, 1995a, 1995b, 1996, 2002, 2007), and at the three Training Institutes[1] of the Open Psychotherapy Centre (OPC)[2]. The above model has been used quite extensively in supervising several kinds of groups (i.e. group analytic, sociotherapeutic, psychodramatic, family and couples' groups) or dyadic meetings (dyadic psychotherapy, psychological assessment, etc.).

Supervision of the psychotherapeutic work is not just a conventional complement of the training process, but a vital element of exercising psychotherapy, almost for life (Kouneli, 2007, p. 59). There is no doubt "that supervision constitutes a decisive field of learning, for any kind of training" (Tsegos, 2002, p. 162). More specifically, the importance of supervision in psychotherapy training is indisputable, as it constitutes a main source of knowledge, especially for new trainees who are confronted with a plethora of theoretical and experiential stimuli, which increase their anxiety, confusion and fear.

1. Conceptual clarifications

Before any further discussion, it would be useful to clarify some concepts and terms regarding supervision. First of all, the term 'supervision' itself derives

---

[1] Institute of Diagnostic Psychology, Institute of Psychodrama-Sociotherapy, Institute of Family Therapy. The corresponding training programmes include personal therapy, theory, clinical practice, communal activities and supervision.

[2] Open Psychotherapy Centre is an autonomous, self-sufficient, non-profit day care unit, not financially supported by any organization inside or outside Greece, founded 30 years ago (1980). The therapy services provided are addressed to individuals who face any type of psychiatric problem on the condition that they are coming on their own free will. The Therapy Department (550 patients per month) includes a great variety of activities as assessment, individual and group psychotherapy (group analysis and group analytic psychodrama), family and marital therapy, children and adolescents therapy and also the autonomously functioning psychotherapeutic communities (daily, fortnight and summer TC). The Training and Research Department includes the seminar's unit (Introductory courses), the training institutes, library, research, publications etc.

from *super-*[3] and *vision*[4]. According to Webster's Dictionary, supervision is "the act of overseeing; inspection; superintendence; oversight". A major distinction should be made regarding supervision in training: instead of a rather unilateral, passive view 'from above', it is more useful to adopt an active, multilateral/manifold seeing. Closer to the second definition is Plato's concept of 'synavgia', which defines *sight* (Tsegos, 2002; Smicht, 2008).

In the Platonic model, the person who sees and the one who is being seen are met in a joint harmonious sight, while the Cartesian model refers to a gnoseology, where there is a dichotomy between subject and object. According to Plato, *sight* is an encounter of two visions: one deriving from the object which sees and the other from the subject which is being seen[5]. Accordingly, we consider the vision of the supervisory group as encountering the vision of the supervised activity.

Supervision is considered as an important source of knowledge (differentiated from information; Tsegos, 2002), as it stands in between theory and practice, allowing the integration of conceptual thinking and experiential knowledge (Behr, 1995, p. 4). According to B. Croce (1909), knowledge can take two forms: it is either intuitive or logical knowledge; knowledge obtained by means of the intellect; knowledge of individuals, or knowledge of the universals; of particular things, or of the relationships between them; it is, in short, either that which produces representations or that which produces concepts. In other words, the procedure of knowledge concerns not only the object (psychotherapeutic approach), but also the subject (in our case the student), and is achieved mainly through experiences, which are influenced by mutual interaction, and also the setting, which has to be appropriate for the desired goal (Tsegos, 2002, p. 162).

---

[3] From L. adverb and preposition *super* "above, over, on the top (of), beyond, besides, in addition to," from PIE base **uper* "over".

[4] From L. *visionem* - nom. *Visio* – "act of seeing, sight, thing seen," from pp. stem of *videre* "to see," from PIE base **weid-* "to know, to see".

[5] Plato decides to deal with cosmos, in one of his last works, Timaeus. The division between the visible world and the transcendental ideas is present, but an effort is made to bridge the gap via triadic relations: the "seeing" and the "seen" need a third reality, light. Light emanates from the eye and meets the external light of the object. Thus subject and object are joined into a common act, the συναύγεια – synavgia (sharing of light): "When the light of day surrounds the stream of vision, then like falls upon like, and they coalesce, and one body is formed by natural affinity in the line of vision, wherever the light that falls from within meets with an external object. And the whole stream of vision, being similarly affected in virtue of similarity, diffuses the motions of what it touches or what touches it over the whole body, until they reach the soul, causing that perception which we call sight" (Plato, Timaeus, 45c).

Without dispute, the contributing factors determining and constituting any training are of course directly analogous to the prevailing philosophy of the institute, which provides the training. Thus, it is essential for every institute to clarify its *basic philosophy*, the goal of the training or *ethos*. Whether hidden or declared, the prevailing targets of any training institute will differ: for example, in a state-supported institution compared to an independent one; or between an established and a new one. Additionally, within any given institute, the objectives of the teachers (i.e. survival, prosperity, and the glory of the institute) are not usually the same as those preoccupying the students (i.e. gathering of information, familiarization with techniques and applications, qualification within the shortest time and with the minimum effort).

These differences are not usually discussed; and when they are, they are frequently dealt with as being 'interpreted'. This is particularly so as one of the professed ideas of psychotherapeutic institutes is achieving insight, which as we believe, should not be the only objective. After all, if "insight without emotional experience is of limited value" (Foulkes, 1969), then this is particularly true for real relationships – and much less for the reductionistic ones ('as if') of psychoanalysis. Insight can be best acquired by exercising the art of relating. There is a way to 'throw a bridge' across, if both sides – trainers and students in this case – are reminded that their basic job is to learn, to teach and mainly to enjoy the art of relating, and this can best be taught and learned by doing it. The training programme is thus based on "ego training in action" (Foulkes, 1964, p. 82 and p. 129) and its objectives are: the acquisition of relevant skills, the acquisition of a clear identity, the consolidation of relevant theory, the learning of role by differentiating it from status and quality, the familiarization in the art of leadership, by differentiating power from strength (Tsegos, 1996). Supervision has a crucial contribution in achieving the above.

The other questionable issue concerns psychotherapy: is it an *art or a science*? Freudian psychoanalysis struggled to prove itself as scientific since its very beginning. New psychotherapeutic approaches (among them group analysis, therapeutic community, psychodrama and family therapy) avoided the above pseudo-dilemma. More specifically, Foulkes considers group analysis as a *therapeutic art*, borrowing from both theoretical and exact sciences, while psychodrama is also called a *healing art.* It is therefore essential to involve in the training, beyond the cognitive element, the emotional and the imaginative as well, in particular, when we deal with a creative and multilevel procedure, such as group psychotherapy.

Finally, we should refer to *mirroring*, which is widely considered as a major therapeutic factor in group psychotherapy. Foulkes describes as 'mirror reactions' or 'mirror phenomena' a set of very important therapeutic factors in an analytic group[6].

> The group situation has been likened to a hall of mirrors where an individual is confronted with various aspects of his social, psychological or body image. By a careful inner assessment of these aspects, he can achieve in time a personal image of himself not grossly out of keeping with the external and objective evaluation. He can discover his real identity and link it up with past identities (Foulkes & Anthony, 1957, p. 150).

Elsewhere he quotes: "It is easier to see the other person's problems than one's own" (Foulkes, 1948, p. 167). The structure of the group is the mirror's setting; its boundaries of time and space are the mirror's frame (Pines, 1984). Not just the patient, who sees himself more and more in and through the group, but groups, can also mirror each other (Terlidou, 1989). This is the case of multiple mirror phenomena, occurring between groups and namely between a therapeutic group and its supervising group.

2. Description of the model: novelties in the structure and the procedure

Group supervision seems to be the most adequate approach in order to moderate the problems arising from the dyadic relationship (issues of roles and relationship between supervisor and supervisee, issues of counter-transference, the issue of the 'unmistakable' supervisor etc.), especially in the training of group therapists. However, group supervision, apart its advantages[7], is not considered as panacea. Problems related to such approach are among the following: dysfunction of the supervisory group (antagonism, distrust, malignant mirroring, obsession to details etc.) and inability for direct and practical solutions.

These are some of the reasons which brought us to experiment with a new model of supervision: a) the anxiety, leading at times to unwillingness to present a group session, particularly among the more inexperienced students, in spite of the supervisor's encouragement and support; anxiety which was connected with the fear of being criticized by the supervisor or their own fellow students, b) a tendency for the supervised students to say the 'right thing', anticipating the

---

[6] Group analytic psychotherapy is not an application of psychoanalysis in the group, but a method and technique based on the dynamics of the group. It is therapy *in* the group, *of* the group (the conductor's included), and *by* the group, the group providing the context in which the individual person is treated (see further information in the work of S.H. Foulkes).

[7] Advantages of supervision in groups are: the culmination of a safe environment, the interaction between fellow trainees, support and the utilization of the supervision group as a mirror of the supervised activity.

preference of the supervisor or imitating the style of the conductor of their personal therapy group, c) an inclination on the student-presenter's part to 'improve', shape or even falsify or 'forget' material in order to avoid criticism, and d) competitive phenomena among students, which inhibited and blocked the supervisory process. As a likely result of all these factors, there was a marked tendency for intellectualization, theoretical discussions and a noticeable dependency on the supervisor (Tsegos, 1995a, p. 120-121). All the above are common phenomena, occurring in group settings, besides the good intensions of the supervisor or the supervisees. The proposed changes concerned both the structure and the procedure of the supervising process.

The structure of the supervisory group: Supervision takes place within a group setting. The supervision group consists of the fellow students – the supervisors- and one experienced observer (who is a member of the training committee or last year's student). A student coordinates the supervising group. It should be stressed that the function of the observer is that of a group member. Of course, he/she is more experienced, but he/she is not there in order to co-ordinate the group.

The supervisory process: During the *presentation* stage, a therapeutic activity is presented, with the student reading out his/her notes and marking group exchanges on the blackboard. The supervising group sits in a semi-circle in front of the blackboard and is expected to take an active part in the second stage of supervision. This is the *analysis*, which follows the presentation stage. During the analysis each of the supervisor-students follows the presentation procedure and records on a special format, the *supervision protocol,* the following:

- His or her *emotions/feelings* during or after the presentation, for example sadness, euphoria, anger, fear, etc.
- His or her *fantasies* (images that passed through the mind during the presentation), for example 'a sinking boat', 'children playing in a schoolyard, an 'ant-nest', etc.
- Which are perceived as the main *themes or topics* of the presented activity?

After the presentation, each of the students reads out his/her notes. All this material is written down on the blackboard.The final stage is the *synthesis.* After all the material of the analysis stage has been recorded on the blackboard, there follows a discussion, chaired by the conductor-student, concerning the interrelation of all material produced, representing the repercussions of the presented session. If there is evidence that the material produced in feelings, fantasies and topics makes sense, the session is completed with relief, along with some suggestions concerning the technique or the way that the therapist or therapists made their interventions.

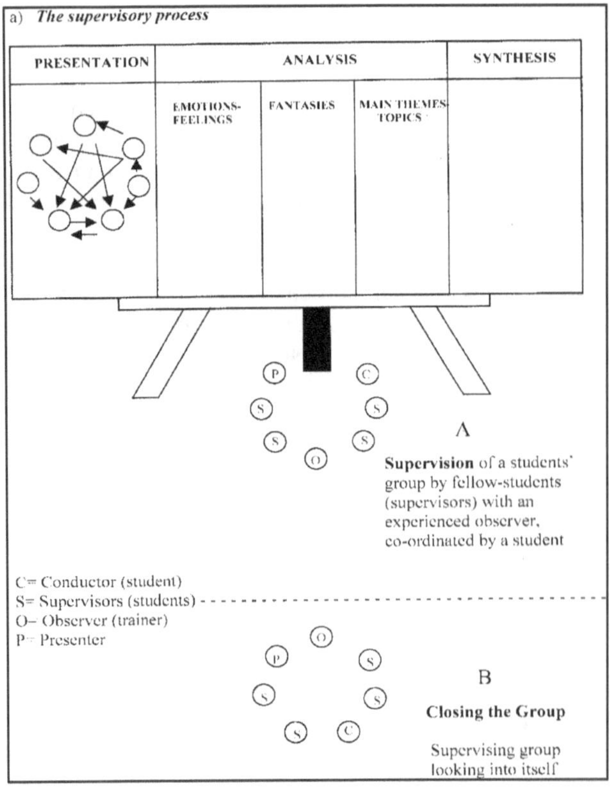

Figure 1: The supervisory process

On the other hand, if there is a lot of irrelevance and incoherence, followed by a tense atmosphere, the group tries to find the reason for this; it may be connected either to the presented material or the observing group, or both. At times the supervising group gives evidence that it is blocked, or that it is not productive, or generally there is something 'going on' with the material that the supervision group offered during the analysis stage. Such occurrences usually arise as a repercussion of mirroring or resonance phenomena coming from the presented group, or because something is indeed 'going on' in the supervising group. It is of course up to the skill of the group to diagnose such a case and intervene accordingly (Tsegos, 1984, 1995a, 2002; Kakouri & Tsegos, 1993; Kouneli, 2005, 2007). In this case the students (members of the supervision group) reform in a circle and face the situation as an affair of the present group.

The supervision protocol: The whole procedure can be further clarified by examining the supervision protocol, a basic structural element, which:

"...not only makes the role of the conductor – who is a student – easier, but structures the whole process itself. The version currently in use serves as a rule of thumb even to very inexperienced students, enabling them to participate in the supervisory procedure right from the beginning actively and effectively. The various parts of the Supervision Protocol also provide important information for dynamic (structural) and practical as well as research purposes" (Tsegos, 1995a, p. 125).

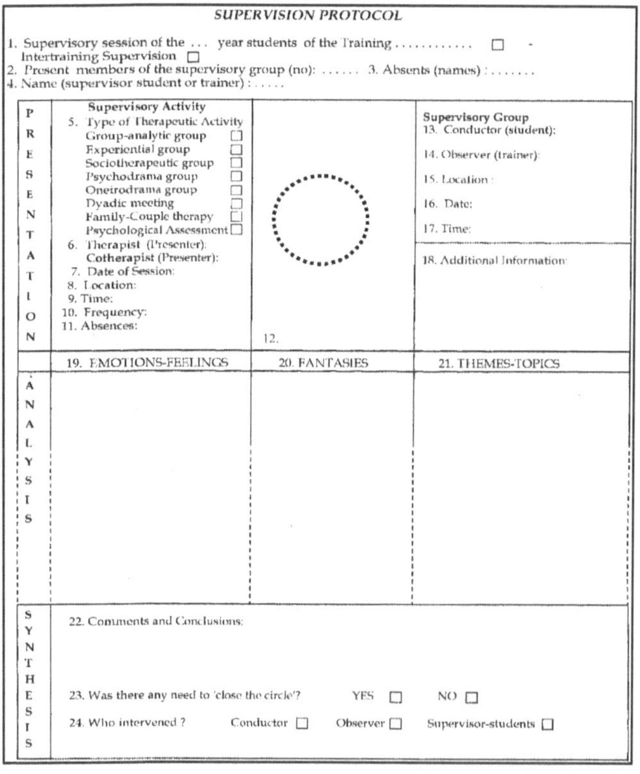

SUPERVISION PROTOCOL

1. Supervisory session of the ... year students of the Training .... ☐ - Intertraining Supervision ☐
2. Present members of the supervisory group (no): ...... 3. Absents (names) : ....
4. Name (supervisor student or trainer) : .....

PRESENTATION

Supervisory Activity
5. Type of Therapeutic Activity
Group-analytic group ☐
Experiential group ☐
Sociotherapeutic group ☐
Psychodrama group ☐
Oneirodrama group ☐
Dyadic meeting ☐
Family-Couple therapy ☐
Psychological Assessment ☐
6. Therapist (Presenter):
Cotherapist (Presenter):
7. Date of Session:
8. Location:
9. Time:
10. Frequency:
11. Absences:

12.

Supervisory Group
13. Conductor (student):
14. Observer (trainer):
15. Location :
16. Date:
17. Time:
18. Additional Information:

ANALYSIS

19. EMOTIONS-FEELINGS | 20. FANTASIES | 21. THEMES-TOPICS

SYNTHESIS

22. Comments and Conclusions:
23. Was there any need to 'close the circle'? YES ☐ NO ☐
24. Who intervened ? Conductor ☐ Observer ☐ Supervisor-students ☐

Figure 2: The supervision protocol

De Mare (1972), Foulkes (1975) and Pines (1983) have stressed the significance of structure, process and content, as a matter of great importance for the function of group analytic situation. Structure and process define the emergence of content. Adequately, the supervision protocol structures the process of the supervisory group; it serves as a means, through which the structure, process and content of the supervised activity becomes visible. This rich and multi-dimensional material can be utilized, for the benefit of the supervised activity.

3. Illustration of a group supervision

The following example presents a group supervision session of a psychodramatic group and the 'mirror phenomena' occurring in different levels, their analysis and interpretation, as well as their effects on many levels of interrelations. The supervisory group, which is running within the context of the training community of the Institute of Psychodrama-Sociotherapy, consists of 6 members (5 trainees and a trainer-observer), its frequency is once a week, and the duration is 2 hours.

The exemplified group analytic psychodrama group is diagnostically heterogeneous (severe personality disorders and psychosis), consists of 7 members and the conductor, a 2nd year student of the Institute of Psychodrama and Sociotherapy. It is the first session that she is the conductor; until now, she was the co-therapist; the previous therapist had left the group due to work.

Psychodrama session (in summary): The group starts discussing on an absent member, without any comment on the farewell of the previous conductor! In the following conversation the theme is the revelation of psychological problems to new acquaintances and more specifically what someone is 'obliged' to reveal. After a while the group members have an argument on whose psychiatric problem is the most serious. The conductor suggests to play, but the group continues talking; the conversation lacks feelings and is full of stereotypical advices and 'philosophical' arguments on mental health. A senior member suggests playing and proposes a theme[8]: 'a rollicking/noisy class in high school'.

During the procedure of sharing the roles, no one is willing to take the role of the teacher. The class therefore is left without a teacher! One of the members

[8] In group analytic psychodrama, the scenario is proposed by any member of the group and the enactment is organised by the whole group. The purpose of that is the full utilisation of the group's therapeutic potential, since we consider that the enactment is not only a representation of the inner conflicts or impulses of one person (the protagonist), but that of the whole group. We assume that, in the first phase, the internal reality of each member is identified with the group matrix. In the second phase it is externalised through dramatic representation (as repetition of action and not real action), and during the third phase it is reconstituted (as a group and individual matrix) and enriched with the experience of action conflict (Papadakis, 1984). The above procedure is essential, as it offers the opportunity for corrective emotional experience in a short period of time. This is especially the case for severely disturbed patients, who have difficulty in expressing themselves in groups, where verbal communication is prominent (Vassiliou et al., 2006).

does not wish to enter the enactment. This is legitimate. Additionally however, he adopts a provocative attitude, by having his back turned to the stage; nobody mentions his behaviour! During the enactment, all the 'students' are united against the absent teacher; they make aggressive, even offensive, comments and plan practical jokes, laughing and having a good time. During the closure phase, all the members avoid discussing on the enactment, opting instead to recount memories from school. The conductor feels uncomfortable and holds back any intervention on the group's attitude. The session ends rather suddenly and the conductor decides to bring it to supervision as soon as possible.

Supervision session: After the presentation of the above therapeutic session (in a more detailed account), the dominant *feelings* reported by the members of the supervision group were anxiety, sadness, anger, tension, insecurity, and abandonment; additionally, somatic reactions were recorded (dizziness, tummy-ache). It is worth mentioning that there was no differentiation of feelings during the presentation; usually during the presentation of the enactment and the closure phase, more positive feelings are provoked, as the group usually achieves a degree of relief or catharsis. The *fantasies* have to do mainly with loneliness, fear and abandonment: 'a girl alone and frightened in a play-yard', 'bees flying around – but no flowers to feed them', etc. The conductor-student of the supervision group says that he had no fantasies; the others comment that this out of the ordinary for him (usually he is full of fantasies) and he responds by saying that the only image stuck in his mind was a 'black screen', 'but this is not a proper fantasy, is it?' The supervision group starts talking about what is proper and what is not. The atmosphere is full of tension and the presenter confesses that she feels shame, supposing that all the feelings of the supervisory group reflect her inability to conduct the therapeutic group properly. The observer reminds them that they should concentrate on the recordings on the blackboard, and not rush to conclusions.

During the stage of *synthesis*, all the feelings and fantasies (the 'black screen' included) were correlated to each other and to *themes/topics* (absences, psychological problems, new acquaintances, 'obligations', old memories). The supervision group realizes that the written data were not only complementary to each other but that they also had a central drift: fear and anger are so intense that they may stop any function (somatic reactions – 'black screen'). On the other hand, the group's feelings (sadness, insecurity, abandonment) and themes (new acquaintances, old memories), indirectly imply a nostalgia and/or a bereavement of the previous group situation, i.e. the previous conductor. The fact that feelings and fantasies did not differentiate from one phase of the psychodrama group to the other was commented on. According to the concept of mirroring, the supervision group recognized the problems and the tension, which dominated the therapeutic group, because all that material was a reflection of its own dynamics:

- the therapeutic group refrain not only from processing its feelings, but even from naming them, throughout the session; it seems that enactment did not lead to relief or catharsis, even though it was so 'cheerful' (adolescents in a high school).
- in the beginning, the supervisory group was overwhelmed by feelings, but through the supervisory process it was able to identify them. Then, even though there was a tendency to avoid the processing of feelings, through speculation ('what is proper', 'I feel responsible'), when focusing on the structural elements of the supervision procedure (to whatever is written on the blackboard), the group became more relaxed and was able to proceed on creative thinking; this was particularly the case for the presenter-student, who realised that she does not have to bear the brunt alone.

The vision of the supervisory group encountered the vision of the supervised activity and had an effect on:

The psychodrama group: the presenter could both recognize and understand the very intense feelings, which were expressed and was gradually able to trust her own self as well as the group that they could bring those feelings to light (and not absorb them like a 'black screen'). It would be more useful to explore these feelings through playing with them, and focus on the enactment and the closure phase (precisely as the supervision group did, focusing on synthesis). The therapeutic group was then able to relate those feelings to the absence of the previous conductor, to the challenge of the new conductor and to the personal unconscious fears and anxieties.

The supervision group: The relationship between the members of the supervision group was strengthened and in the following sessions the group was very productive, with a great deal of important benefit for the group work. The reflection, through the mirror of the therapeutic group, of the supervising group situation helped all the members (supervising students) to recognize and accept their own feelings and trust even more the supervisory procedure.

Through the two procedures (supervised activity and supervision by peers) "the trainee can be convinced that knowledge, and the enterprise of transforming unconscious material into conscious, can be better achieved through relating" (Tsegos & Tseberlidou, 2002, p. 251). After all, the concept "con-scious" in its Latin source meant "to know with" (to share knowledge with another) and not to "know in oneself alone" (Whyte, 1960, p. 43).

4. Research study on group analytic supervision

It seems rather strange that there is a lack of research evidence concerning the different models of supervision (dyadic/group) or the ways through which a better combination between theory and practice would be succeeded for the trainees. The lack of relevant studies concerning group supervision in the existing bibliography, prompted us to research the issue in depth.

I. First of all, the issue of *the significance of supervision* as an integral part of training. That is, which part of the training (personal therapy, theory, clinical practice, supervision) is evaluated, among the students, as the most important and which among the trainers. According to the findings of a 'Research study on the training factors in a training community'[9] (Fikiori, 2007), the majority of the participants (trainers and trainees) evaluate personal therapy, clinical practice and supervision (in that specific order) as the most important factors for personal development and establishment of the professional identity, while supervision is the most important factor in helping trainees to fulfil the demands of clinical practice (83%) and compose theoretical knowledge and clinical experience (69%).

Additionally, through open questions, the trainees recorded that the most significant changes which occurred during training were improvement in the relations with others, flexibility in the relations to others, interchange of roles, improvement of the self-image, empowerment of the ego strength, improvement in the expression of feelings, self-disclosure, self-boundaries, acceptance of critics etc. Again, personal therapy, clinical practice and supervision were indicated as the place where the above changes mostly occur, but with a new entry, that of communal training activities[10] on the second place.

II. Secondly, the fundamental question of which is *the element that the supervision reflects or mirrors*: the supervisor, the supervisee, the supervised activity or the supervisory group. The findings of an extended research, '*Group analytic supervision: research study on protocols derived from group supervision activities*' (Karayanni, 2004; Tsegos et al., 2004)[11], indicate considerable

---

[9] The aforementioned study, through a structured questionnaire, was addressed to the students and the trainers (N=74) of the four institutes (group analysis, diagnostic psychology, psychodrama - sociotherapy, family therapy) with the purpose of evaluating the impact of the different training activities in two major areas: personal development of the trainees and establishment of the professional identity.

[10] The entire training is materialized through a communal scheme, a training community, where trainers and trainees work together and where common activities contribute so that both sides derive maximum benefit in their pursuit of knowledge (Tsegos, 1999).

[11] *Methodology:* Record of the data derives from the archives of the supervision protocols of all supervised activities of a decade (1992-2002), as the model had been established and the participants had been fa-

differences in each therapeutic activity per se, though the supervising population is mixed (the participating students-supervisors derive from all four institutes, i.e. group analysis, diagnostic psychology, psychodrama-sociotherapy, family therapy). It seems that the supervising group echoes and mirrors each therapeutic activity correspondingly. Summarizing, the main findings are:

- The highest means of *positive emotions/feelings* and *fantasies* occurs in therapeutic communities groups and group analytic psychodrama.
- The highest means of *negative emotions/feelings* and *fantasies* occurs in family therapy, psychological assessment, couples' therapy and dyadic psychotherapy.
- It seems that while most *group* activities present the highest means of positive *emotions/feelings* and *fantasies*, the *dyadic* activities such as psychological assessment, dyadic psychotherapy present the highest means of *negative emotions/feelings* and also *negative fantasies*. Additionally, family/couple activities (which are somehow grouped) present a high means of negative emotions/feelings and fantasies.
- All the supervised activities present almost the same means of *themes and topics*, with a small deviation.

In conclusion, the tendencies, which appear in the supervisory protocols of the different therapeutic activities, both quantitative and qualitative, are common. Since this occurs regardless of the synthesis of the supervising group, we presume that the group reactions are mainly related with the supervised activity. Therefore, any critical observations regarding group supervision (i.e. development of group phenomena that prevent or distort the supervisory process) do not seem to stand valid concerning the specific model of supervision. By no means do we ignore the existence of other dynamic factors, which are related to the developed relationships among the participants of the supervisory group and they may influence the supervisory process. And because these factors may prevent and complicate the supervisory group process, the 'safety valve' of the Greek model for such phenomena is to 'close the circle', that is, to momentarily

miliarised with the process. *Sample of the study:* The total number of the supervision protocols was 225, corresponding to 1027 different entries of supervising students. *Categorization of the data*: The entry and analysis of the data was divided in three categories: the first one was concerning emotions/feelings, the second fantasies and the third one themes and topics. Given the great number and variety of entries, we decided to apply the following categorisation: a) Positive emotions/feelings (i.e. pleasure, familiarity, joy etc.), negative emotions/feelings (i.e. agony, anxiety, sorrow etc.), indeterminable emotions/feelings (i.e. surprise, query, astonishment etc.), b) Positive fantasies (i.e. children playing happily), negative (i.e. a monster threatening a young couple), indeterminable (a small boat sailing in the winter), c) the third category, which was main themes and topics, was classified according to the number of entries for each protocol and the number of resembling entries for each protocol.

leave its supervising task and become engaged in issues relating to itself and to the relations among its members.

The above findings are important in relation to the study of the specific model of supervision; however it is necessary to clarify that they are not connected to any therapeutic outcome. The fact that certain therapeutic activities cause more positive or negative reactions to the supervisory group, does not mean that they are more or less therapeutic. On the contrary, it is the different character of these activities, which is distinguished, a fact which may constitute the motive for future interesting research studies.

III. The above study revealed certain methodological difficulties in categorising *emotions/feelings* and *fantasies* and led to new studies. Even though emotions/feelings is a common place in psychotherapy, there is no evidence on the subject; i.e. in which way the supervision procedure bring to prominence emotions/feelings and utilizes them. According to the findings of the study 'Emotions in group analytic supervision' (Mitroutsikou, 2005), there are different kinds of entries under the section *feelings/emotions*:

- entries which purely describe an emotional state (i.e. anger, pleasure, sorrow),
- entries which describe the perception of an event in the supervised activity, on behalf of the participants (i.e. relief, tension),
- entries which describe a somatic reaction of the participants (i.e. headache, exhaustion),
- entries which describe an event by interpreting it (instead of writing about a fight, the student mentions it as a competition),
- entries which describe a condition (i.e. perplexity, interest).

It is affirmed, once again, that while we constantly deal with emotional matters in our clinical work, it is rather difficult to deal with our own feelings/emotions and there is a tendency to interpret rather than to feel. This is critical in the following sense: in the specific model of supervision (which 'obliges' the participants to clarify and write down their emotional state, as a part of a structured procedure) some difficulties are observed; one could therefore only imagine what could happen when such a procedure is absent. It is a great advantage for the students to have to record their feelings from the very first day in their training. However, the above findings indicate that the Greek model of supervision is not a panacea for the latent, 'psychoanalytic' or authoritarian tendencies of the 'presumptive psychotherapists' (trainees) or the 'highly regarded' trainers. The specific model can be more productive if the experienced observer is convinced of its usefulness, and the context (institution) within which it takes place trusts the group more than the experts (Tsegos, 1995a, p. 128).

IV. The imaginative level is of great importance in every human activity, all the more in psychotherapy. So, it is rather strange, that even though we utilize and enhance imagination for therapeutic purposes, there is a lack of evidence on how supervision turns to advantage the produced imaginative material. A research study on 'Fantasies in group analytic supervision'[12] (Panagopoulou, 2008; Zerva, 2008) produced the following findings: following the determinant factors of 'Rorchach and Rorchach Interaction Scale', an attempt was made to correlate the fantasies, produced by the supervisory group, with the type of the presented material. Quantitative data shows differences and similarities in qualitative characteristics of fantasies, according the type of the supervised activity. Concisely:

- A common feature is that the supervision group functions on a high level of creativity, internalised thinking, an ability for empathy, an intense interest for people, as well as a tendency for exercising exceeding critique.
- Every type of supervised activity provokes a different kind of reaction (imaginative and emotional).
- The differences, among others, are: all supervised activities, except psychological assessment and dyadic psychotherapy, provoke to the supervisory group tensions or/and confrontation. Dyadic meetings and family/couple therapy provoke a stirring of personal inner conflicts. Group activities (group analytic groups, group analytic psychodrama, therapeutic community groups etc.) impel supervisory group to function rather emotionally, in comparison to dyadic activities, which lead to rational thinking.

5. Concluding Remarks

It appears that the Greek model of supervision contribute a lot towards creating a relaxed atmosphere and reduces the hesitation of inexperienced students to present their group. This is due to the minimization of the inhibitions created by superegotic situations, such as the interventions of one and only supervisor-trainee. Additionally, the tendencies towards authority and authenticity are lessened, both for the trainers and the students. The group holds an important role in the enhancement or the weakening of the roles of the expert or the beginner. What Foulkes writes about the conductor is also true for the role of a trainer in our case: "The conductor must avoid becoming too important and must keep to

[12] The sample of the study (1320 entries under the section *fantasies)*, derived from supervision protocols from different supervised activities (group-analytic groups, therapeutic community groups, group-analytic psychodrama, couple therapy, counselling for parents, family therapy, psychological assessment).

the background... The group will learn to rely more on itself and be correspondingly of the truth of its own findings" (Foulkes, 1975, p. 111).

The members of the supervisory group participate actively in the procedure, in multiple levels: free intellectual function (record of the topics of the session), emotional function (insight and record of personal emotions/feelings) and imaginative function (personal fantasies). Trainees, are thus exercised not only in avoiding to take cover behind the role of the detached expert, but also in gaining awareness of their inner processes, which are provoked by their clinical work. The way in which the experienced observer is functioning (more horizontal) reinforces the trust towards the group function (Tsegos, 2002).

The specific model provides an opportunity for maturation and development, both in personal and professional level through the constant interchange of roles (supervisor, supervisee, conductor), the familiarization with group phenomena (mirroring, resonance, condenser, etc.) and the differentiation between *leading* and *co-ordinating*, as well as between *status* and *role*, and therefore between concepts such as *authoritative* and *authoritarian*. There is no doubt that all these concepts and roles can also be taught through the process of internalisation, by means of observing seniors. However the enacting or playing these roles by oneself or by frequently observing peers carrying them out, adds a memorable experience (Tsegos, 1993).

References

Behr, L. (1995). The Integration of Theory and Practice. In M. Sharpe (Ed.), *The Third Eye. Supervision of Analytic Groups* (pp. 4-17). London: Routledge.

Croce, B. (1909). *Aesthetic as Science of Expression and General Linguistic.* BiblioBazaar (2006).

De Mare, P.B. (1972). *Perspectives in Group Psychotherapy. A Theoretical Background.* London: Allen and Unwin.

Fikiori, I. (2007). *Evaluation of the Training Factors in a Training Community. Research Study.* Dissertation. Institute of Psychodrama and Sociotherapy, Open Psychotherapy Centre, Athens.

Foulkes, S. H. (1948). *Introduction to Group Analytic Psychotherapy. Studies in the Social Integration of Individuals and Groups.* London: Heinemann (Reprinted, London, Karnac Books, 1983).

Foulkes, S. H., & Anthony, E. J. (1957). *Group Psychotherapy. The Psychoanalytical Approach.* London: Maresfield Reprints (reprinted 1984).

Foulkes, S. H. (1964). *Therapeutic Group Analysis.* London: Maresfield Reprints.

Foulkes, S. H. (1969). Summary and Conclusions. In S.H. Foulkes & G. St. Prince, *Psychiatry in a Changing Society.* London: Tavistock Publications.

Foulkes, S. H. (1975). *Group Analytic Psychotherapy. Method and Principles.* London: Gordon and Breach (Reprinted, London, Karnac Books, 1986).

Kakouri, A., & Tsegos, I.K. (1993). Boundaries and Barriers in Peer Supervision. In W. Knauss, & U. Keller (Eds.) *Proceedings of the 9th European Symposium in Group Analysis.* Heidelberg: Mattes-Verlag.

Karayanni, V. (2004). *Group Analytic Supervision. Research Study on Protocols Derived from Group Supervision Activities.* Dissertation. Institute of Group Analysis, Athens.

Kouneli, E. (2005). The Image of the Self in the Mirror of the Supervising Group. *Group Analysis*, 38(4), 558–568.

Kouneli, E. (2007). Group Analytic, Communal Supervision. In I. K. Tsegos & Collaborators. *Open Psychotherapy Centre (1980-2007).* Athens: Enallaktikes Ekdoseis (in greek).

Mitroutsikou, E. (2005). Emotions in Group Analytic Supervision. Dissertation. Institute of Psychodrama and Sociotherapy, Open Psychotherapy Centre, Athens.

Panagopoulou, T. (2008). *Fantasies in Group Analytic Supervision.* Dissertation. Institute of Family Therapy, Open Psychotherapy Centre, Athens.

Papadakis, Th. (1984). *Group Analytic Psychodrama.* Dissertation. Institute of Group Analysis, Athens.

Pines, M. (Ed.) (1983). *The Evolution of Group Analysis.* London: Routledge.

Pines, M. (1984). Mirroring in Group Analysis as a Developmental and Therapeutic Process. In T. E. Lear (Ed.), *Spheres of Group Analysis.* Group Analytic Society Publications.

Plato (1965). *Timaeus and Critias.* Penguin Books.

Smicht, B. (2009). *The Concept of Mirroring from Ancient Greek Literature to Καθ' Ἡμᾶς Group Analysis.* Dissertation. Institute of Family Therapy, Open Psychotherapy Centre, Athens.

Terlidou, C. (1989). *The Group Analytic Therapeutic Factors in the Psychotherapeutic Community of the Open Psychotherapy Centre.* Dissertation. Institute of Psychodrama – Sociotherapy, Open Psychotherapy Centre, Athens.

Tsegos, I. K. (1984). Experimenting on Group Analytic Supervision. 6th European Symposium in Group Analysis, Zagreb.

Tsegos, I. K. (1993). Strength, Power and Group Analysis. *Group Analysis*, 26 (2), 131-137.

Tsegos, I. K. (1995a). A Greek Model of Supervision: the Matrix as Supervisor – a Version of Peer Supervision developed at I.G.A. (Athens). In M. Sharpe (Ed.), *The Third Eye. Supervision of Analytic Groups.* London and New York: Routledge.

Tsegos, I. K. (1995b). Further Thoughts on Group Analytic Training. *Group Analysis,* 28 (3), 313-326.

Tsegos, I. K. (1996). Small, Medium and Large Groups in Group Analytic Training. The Training Community of I.G.A. Athens. *Proceedings of the Egatins' Study Days.* Ljubljana: Egatin.

Tsegos, I. K. (1999). Training. Establishing a Professional Identity. In P. Campling & R. Haigh (Eds.), *Therapeutic Communities. Past Present and Future* (pp. 189-206). London: Jessica Kingsley Publishers.

Tsegos, I. K. (2002). *The Disguises of the Psychotherapist.* Athens: Stigmi (in greek).

Tsegos, I. K. & Tseberlidou, M. (2002). The Oneirodrama Group. The Therapeutic and Supervisory Process of a Dream Drama Group. In C. Neri, M. Pines & R. Friedman (Eds.), *Dreams in Group Psychotherapy. Theory and Technique* (pp. 233-253). London: Jessica Kingsley.

Tsegos, I. K., Karayanni, V., Karapostoli, N., Morarou, E. (2004). An Extensive Outcome of a New Research to Supervision: A Research Study Derived from Supervision Protocols of Students' Psychotherapeutic and Psychometric Activities. *The European Journal of Psychiatry,* 18, 61-70.

Tsegos, I. K. & Collaborators (2007). *Open Psychotherapy Centre (1980-2007).* Athens: Enallaktikes Ekdoseis (in greek).

Vassiliou, O., Livas, D., Karapostoli, N., & Papadakis, Th. (2006). Theoretical Developments and Various Applications of Group-Analytic Psychodrama. *International Journal of Therapeutic Communities,* 27 (2), 275-290.

Whyte, L. L. (1960). *The Unconscious Before Freud.* London: The Estate of Lancelot Law Whyte (Greek translation, 2010, Athens: Enallaktikes Ekdoseis).

Zerva, E. (2008). *Fantasies in Group Analytic Supervision.* Dissertation. Institute of Diagnostic Psychology, Open Psychotherapy Centre, Athens.

Supervision and evaluation: objectives, practices and helpful aspects

Hannes Krall & Jutta Fürst

Supervision in psychodrama training aims to improve the learning of trainees through reflection on their own practice. The deepening of theoretical knowledge, development of technical skills and the focusing on personal and interpersonal awareness should be taken into consideration.

Although there is a broad consensus that supervision of trainees is crucial for their professional development, there is not much attention given to how the process of supervision should be evaluated. Therefore the following chapter focuses on the role of supervision in psychodrama training. It will describe and reflect systematically on the supervision process. It proposes methods of evaluating the outcomes of supervision and investigates how helpful aspects of psychodramatic supervision relate to each other. At the beginning of this chapter the general structure and setting up of supervision will be explained. In a second step the process of supervision will be described along with practical examples and theoretical considerations. In a third step consideration will be given to the results of a pilot study to evaluate outcomes and helpful aspects in the supervision process.

1. Supervision and learning in psychodrama

The word supervision sometimes evokes ideas that there is someone outside who has a better view of the therapeutic process than the supervisee. The supervisor then becomes regarded as a super-mind with outstanding knowledge and wisdom as it is described in the following story:

One day a dervish teacher was walking along a lake meditating and absorbed in his thoughts. Suddenly he was interrupted by a loud call. Obviously a dervish student being on an island in the middle of the lake was intoning the call of the dervish. Performing it perfectly it would allow one to walk over the water. The teacher thought: 'The way he is doing it does not make any sense. It is my duty as an expert to teach this unlucky fellow to do it better.'

The dervish took a boat, went over the lake and did his best to correct the wrong sound of the young ambitious man who was very thankful that the wise dervish gave up his time to teach him.

A short time later the dervish was still in his boat on the way back when, he heard again the voice of the student. It was still the same wrong sound that came from the island. While thinking about how untalented or maybe unwilling some people are he saw the student beside his boat. He had walked over the water coming to the boat of the dervish teacher and asked in a humble voice: "Master, please don't be annoyed. Would you be so kind to repeat the call again? I still have difficulties performing it correctly.

Even experienced trainers have to acknowledge that trainees learn to practice psychodrama in their own way. Sometimes they 'walk over the water' even better than their dervish teacher, who still needs a boat. This story should not diminish the importance of supervision but emphasize the fact, that we still do not know a lot about how to teach this magic art of psychodrama through supervision. We should ask more about objectives and helpful aspects of this learning process. Questions are to be answered like: Which aspects of the supervision are considered as helpful? Are there some techniques that are more effective than others? Does it depend on the trainee which technique is helpful or are there some aspects which are in general more effective than others? If we think as psychodramatists that action and role reversal are more effective than simple verbal reflections, is it possible to verify this presumption also for supervision?

There are only a few studies existing about effects of supervision in training. Some of them describe even negative effects or focus on counterproductive events (Gray, Ladany, Nicholas, & Walker, 2001) or failures of the supervisor (Ruskin, 1994). Others investigate conflictual supervisory relationships (Nelson & Friedlander, 2001). Therefore it makes sense to look closer to the objectives, processes, ways and outcomes of supervision in training.

2. Supervision in the psychodrama curriculum

Supervision is part of the psychodrama training programme at the Institute of Psychosocial Intervention and Communication Research at the Faculty of Educational Sciences at the University of Innsbruck in Austria. The programme itself has to be in accordance with the Austrian psychotherapy law. An advisory board of the Ministry of Health is in charge of accepting and controlling the institutes and their programmes. Trainees have to practise 600 hours of therapeutic work with clients in groups and individual sessions. Each session has to be recorded thoroughly. The number of hours is equal for all psychotherapy training institutes irrespective of the method. Also a minimum of 150 hours of supervision are stipulated by the government. The format and setting that is used for the supervision can be decided by each training institute itself.

According to the psychotherapy law a student must have a minimum age of 24 years and has to be a graduate of a general psychotherapeutic programme called 'Propädeutikum' to start psychotherapy training. The average age of our trainees is 39 years with a range from 27 to 54. The average number of participants in a training group is 14. It ranges from at least 12 to 16 at most. Only 10-15% of the trainees are male. The majority of the trainees (60-75%) are graduated psychologists or pedagogues. From the very beginning of training most of the trainees are already working in social services such as psychiatric hospitals, paediatrics, intensive care, residential drug withdrawal, child protection centres, psychological counselling services, residential institutions for children and adolescents, refugee care services or institutes for mentally handicapped people.

In the third and fourth year of the training the trainees work psychotherapeutically with clients under supervision. Before getting this status they have to complete 240 hours in a self-exploration group and at least 30 hours of individual psychodrama. They also have to pass 210 hours of theory seminars. In addition they have to attend at least 275 hours internship in a health care facility.

Supervision takes place at the university in small groups of 3-5 supervisees. It is offered every 2-4 weeks and has an average duration of 5 working hours (45 minutes). The groups are in general constant regarding their members and change only occasionally.

3. Trainees and their roles in psychodrama training

In the last two decades psychodrama training has become more structured and the content more elaborate. Psychodrama trainers and trainees have become aware that continuous development in theory and practice is essential for the future of psychodrama. It is not enough just to learn methods and techniques in psychodrama. Trainees have to become acquainted with theoretical reflection and research. Having a theory in mind and looking on an interaction from outside helps to structure and to find solutions for challenging situations.

The training programme at the University of Innsbruck aims at supporting the trainee to develop certain roles which are assumed to be important for the work of a psychodrama psychotherapist:

• *The role of the client and protagonist:* As a protagonist the trainee can explore significant events in the past which have influenced his or her behaviour, feelings and thinking in the past and present. Trainees start similar to a client in the role of someone who wants to learn more about his or her own possibilities to overcome difficulties in life.
• *The role of the auxiliary:* The trainee learns to be an empathic double and to be a flexible and helpful group member by taking the role of an antagonist when he or she is chosen for a role in the play of the protagonist. During the integration phase the trainee learns to distinguish between sharing, role-feedback and identification-feedback and to find the appropriate words and sentences for one's feelings and experiences.
• *The role of an observer:* In the training group there are also phases where the trainee is neither chosen as protagonist nor as an auxiliary and therefore is observing the process. It allows also learning by watching the trainers as a proper role model when they are working with a protagonist.

These three roles are mainly developed in the first two years of the training when the trainee is part of the basic ongoing self-exploring training group. Later on trainees develop more advanced roles:

• *The role of the director:* In the advanced training group the analytic and technical aspects are focused on. Trainees are learning how to make a proper warm up, to choose the protagonist and the topic of the group, to set up a scene, to apply the techniques in a helpful manner, to make a good closure and to lead attentively through the reflection phase. In this advanced ongoing group they develop the role of a group facilitator, a creative producer, a helpful healer and a smart analyst (Kellermann, 1972). The trainees are there introduced into the role of the director.
• *The role of the analyser and concept manager:* Beside these two ongoing groups the trainees acquire theoretical and practical knowledge by attending theory seminars which enable the trainees to make a link between their practical experiences and theory (philosophy, concepts of mental health and illness, diagnosis, role pathology, etc.)
• *The reflecting role or the role of the 'inner' supervisor:* In supervision the trainees are expected to develop the ability to look at their therapeutic work from outside comparable with the mirror technique in a psychodramatic enactment. They learn to get distance and to overview what is happening. They can look at the way they act as director or at the relationship between client and director and they can develop ideas and concepts how to improve the situation. The development of this role starts during the processing part of the basic self-

exploration group, is continued in the advanced group and later on in the supervision group. Step by step they develop the role of the 'inner' supervisor of their own work.

For a psychodrama therapist it is important to be able to switch from the role of an empathic double, to an attentive observer, to a clear analyser and thinker and to the role of the director who is able to set clear interventions. Therefore students must develop competences in each of these roles and to switch quickly between them during the supervision process.

4. Methods and techniques in supervision

Experienced practitioners who are attending a supervision group usually have already developed self-confidence in their own professional work and therefore it is easier to them to focus on specific issues of their work.

Trainees instead usually have more doubts than certainties regarding what they are doing – especially when they start to do psychodrama under supervision. They also might be afraid of being assessed by the supervisor, even though supervision is not part of the evaluation process. In this context they might feel inferior to the supervisor with respect to their knowledge and skills in psychodrama.

Various studies report that the relationship to a respectful, supportive and collegial supervisor (Gandolfo & Brown, 1987; Worthen & Mc Neill, 1996) is an important precondition for the disclosure and self-efficacy of a trainee. Reising and Daniels (1983) found out that the anxiety level and dependency on the supervisor is on the beginner's level higher and that they are less ready for confrontation than on an advanced level of practice. Since trainees do not rely so much on their knowledge and competencies, their intuition and feelings regarding their work with the clients, one main task of supervision in training is to assure and strengthen the trainees in their work.

Supervisees sometimes have a lot of difficulties in the beginning to tell what they feel when they are confronted with a problem. They are very quick with interpretations and it needs some time to distinguish between feelings, associations and interpretations although they are good in giving feedbacks after a role play. The goal in this sequence is to learn to appreciate the reflection of feelings as an important source for getting insight about what is going on. In this part the supervisor is more in the role of a teacher. Ronnestadt and Skovholt wrote:

The theory-practice gulf so intensively experienced by the beginning student makes this a 'teachable moment' for focusing on specific counselling skills. The effective supervisor at this level often assumes the role of a teacher, a role different and more limited than at later stages. The student is typically eager to learn specific skills, and supervisors who emphasize this are highly valued. Supervisors who do not have this skill focus are generally criticized and often disappoint their students (Ronnestadt & Skovholt, 1993, p. 397).

4.1 Warm up and creating a safe place

For the supervisor it is important to be aware of the challenges trainees are facing not only in their psychotherapeutic practice, but also in the context of supervision itself. It has to be emphasised that learning processes and creative work of trainees needs a relaxed feeling. Thus, the warm up in supervision can be understood literally to take care and to create a warm atmosphere where everyone feels free to explore without fear. "The supervision space must be able to contain storms, distress, despair at humanity and oneself, hard-headed planning, and tough mutual evaluation" (Williams, 1995, p. 19). The warm up should on the one hand help to create a trustful and secure atmosphere and on the other hand prepare the group for the case work.

Scene 1: *The group – Anna, Magda, Paula and Martin – is meeting for the third time. The supervisees welcome each other and seem to be happy to meet. In the first few minutes the supervisees are sharing personal feelings and events of their private life. The supervisor invites them to walk through the room and to think about enjoyable and difficult scenes of their professional life. Then they are asked to focus on those troubles which weighed them down or puzzled them most. The supervisees choose then some cloths or available objects in the room to symbolise the problems they found.*

Martin folds a green cloth carefully in a square and put something underneath that can only be guessed because the cloth does not lie flat on the floor. A heap made of a black entangled ribbon is formed by Anna. Magda arranges different cloths with strong clear colours like an abstract cubistic painting and beside that a woollen red cloth under a chair. Paula laid a long light blue silken scarf on the floor which is strangulated in the middle by a red string. After that the supervisees are invited to walk between the objects like being in an art exhibition and to tell which feelings (not interpretations) they have when they look at a certain object.

Everyone is asked then how they felt when they heard what the colleagues associated to their symbolic representation of the problem. Most of the feelings fitted to their own, some were different and made them think. After that they are asked to tell what they associate with their symbols. Then the trainees choose sociometrically one piece which they are interested most in this moment. It is Paula's blue silken scarf. And Paula starts to tell her case…

By warming up for the supervision with a kind of relaxation it helps to focus on topics which really matter to them. Especially at the beginning of their training when they do not know what will happen in the supervision they are afraid to be criticised. Depending on their self-esteem they might choose sometimes a case that went rather well. Thus, it seems to be helpful at the beginning to give a voice to their feelings, in order not to feel threatened by questions or by being forced to explain something clearly. When the supervisees choose one case of a colleague not knowing what it is about they follow their feelings and learn to trust them.

4.2 Introducing the case and choosing a supervisee-director

After the case was chosen by the group the supervisee starts to tell in a narrative way about her work. Then the supervisee is asked to put a question she is struggling with. The other supervisee's task is to listen carefully and to be aware in which part of the story they become emotionally involved, which feelings and associations come up during the presentation of the case.

Scene 2: *Martin tells about his client Lila, who has been in therapy for several months. She suffers from an untreatable skin disease that started some months ago and defaces her completely. She feels not loveable anymore and can hardly cope with the idea to show her body again jauntily in a public space. During the therapeutic work it turned out that she has certain sexual desires her partner does not fulfil. She wants him to be a macho man who takes her without asking. In reality her partner instead is someone who is very tender and tentative.*

Martin is not sure how to proceed in this case. He does not understand the connection between Lila's symptom and that strange fantasy.

The supervisor asks the other supervisees what they feel when they listen to this story. The three female supervisees identify a lot with Martin's client and they feel torn by the wish to live in an uncontrolled and spontaneous way and the fear of doing it. The supervisor asks: 'If you were the supervisor of Martin how would you work psychodramatically'. Several ideas how to proceed with the client were mentioned.

The supervisor then asks Martin which idea seems to be most appropriate to him. Martin chooses Anna's plan. Anna proposes to give Martin the possibility to explore the roles of Lila and the macho man in her fantasy scene. Martin should play both parts in order to understand better what is going on. In the supervisory work Anna is then taking the role of the director and goes on stage with Martin to enact the scene. Anna's hypothesis is that there are also fears inside Lila and that the symptom protects her to live the sexual fantasies.

The supervisee-director Anna was chosen by Martin and by that she experiences the appreciation for her idea. Taking the role of the director the supervisee has to change quickly from one role to the other. In order to feel more secure in the role of the director some aid is provided. The supervisees can choose what they find helpful for them when they are working on stage:

- a freezing spray to freeze the scene when the supervisee-director needs help from the group or the supervisor;
- a defroster spray to continue the work on stage;
- a soliloquy of the supervisee-director to let the group know what he or she is thinking about;
- doubling of the director by the supervisor or a group member.

The supervisees are asked about their thoughts regarding the presented case to develop hypothesis and plausible explanations for the symptoms of the client. Caspar (2005) outlines the importance of the competence of information processing of psychotherapists after analysing many studies in this field. Supervisees in a group share their implicit concepts, which are the basis of understanding and planning of interventions.

"Concept competence" of the therapist can be considered as a substantial factor of efficacy of therapy (Kahl-Popp, 2004; Crits-Christoph et al., 1988). It is defined as the complex comprehension of another person which includes an explanatory model/ aetiology and a plan of steps of interventions. A study of Kraler and Andreatta (2006) evaluated the concept competence of psychodrama trainees which was operationalized as the degree of complexity in structure and content in building hypothesis for psychotherapeutic cases. It was verified that the differentiation of the concepts improved a lot from the beginning of the training to the end.

4.3 Specific interventions in supervisory case work

Focusing on a key sentence or a metaphor, looking at the relationship between a presented case work and the biographical experiences of the trainee, deepening the theoretical understanding of clients symptoms or reflecting upon the rational of interventions are some possibilities for specific approaches in supervision. The following examples illustrate some specific interventions of supervisory work.

Focusing on a key sentence

Scene 3: Sonja comes completely exhausted, tired and angry to the supervision group. She is working in a residential institution for young people. It is hard to cope every day with the changing mood of the adolescents swapping between aggression and depression and after having reached a kind of success in stabilising them falling with them into a deep hole when the authorities decides that they have to leave the place or must go back to their parents who treated them badly. Sonja feels that all her work is senseless.

Asking Sonja who is telling her what makes sense in her work or life she says it is an inner voice. The voice is personalised as an old wise woman. The supervisor puts a chair for the old wise woman on stage. Asking the old woman if her work makes any sense the supervisee answers after a role reversal: 'You cannot save the world, be humble and do what you can do!' Sonja feels relieved. The sentence becomes essential for her work.

Focusing on a methodological question

Scene 4: *The topic of the group turns out to be how to finish a therapy. The supervisees share some techniques they have already used. Lisa likes the exercise with two boxes most, where the client can put all good things he or she has experienced into one, and all negative things into the other. She uses it for her client, a man who started to become depressed after the death of his mother. He is idealising his mother a lot. In some session ago he realised for the first time that there were also a lot of situations when he hated his mother. At the end of the therapy Lisa wants to offer him four boxes made of paper by her. One box is for all the good situations and feelings he experienced with his mother, the second is for all the bad situations. The third box is for all the good things he experi-*

enced during the therapy and the fourth box is for the things he did not like in therapy.

Focusing on a metaphor and on the biography of the supervisee

Scene 5: *When Martin is describing his depressed client in his group, everyone gets the image she is like a tree with broken branches. He does not know how to continue with his client. He chooses Anna for being the supervisee-director. She invites him to take on the role of this tree with broken branches and asks him what he needs. He answers that he does not know if he could ever grow again because he is terribly injured. The supervisee in the role of the director seems to get stuck and is telling in a soliloquy: 'I feel scared by this image and don't know what to do. I have no idea now.*

'The supervisor doubles the supervisee-director: 'I thought that going into the metaphor of that tree would help to develop other ideas but now I am in the same situation as my colleague was before. But I can ask Martin if he feels, sees or hears something as a tree – maybe there is a ground or a sun or grass…'

Anna asks then Martin in the role of the tree if there is something around what he can see. Martin discovers a gardener. He chooses Paula to be the gardener, who is described as an experienced man, strong and wise. Anna let Martin reverse into the role of the gardener. In this role he put some healing cream on the stumps where the branches had been. Then he investigates the roots. In the role of the tree again he feels very comfortable and calm because he knows the gardener is patient enough to wait until new leaves are growing. Suddenly Martin remembers that he was always forced by his parents to be more developed then his peers. Back in the role of the gardener he takes on the role of a good and patient gardener.

4.4 Integration and reflection

Discharging the supervisee-director and integration is important after supervisory case work. The supervisor asks the supervisee-protagonist and his director to clean the stage. After that the supervisee-director, the protagonist and the auxiliary egos are giving their role feedback.

Scene 6: *Anna asks Paula and Martin how they felt in the role of the tree and the gardener. Paula says that she was very sad when she saw the injured tree and was bond by thinking about what had happened to this tree. In the role of the tree she felt safe because there was a good gardener there. Anna asks then Martin if he found an answer for his question. He feels very satisfied and calm taking the metaphor of the tree and the patient gardener with him.*

After Martin, Paula and Anna step out of their roles the supervisor asks Martin how he felt accompanied by Anna as a director. He mentions that he learned a lot because Anna was such a patient director who gives him so much time to feel and think. Then the supervisor asks Anna what she got out of her role as director. She shares that she is pleased to hear that Martin appreciated her patience which she always referred to as slowness.

After that feedback the supervisor asks the group members to arrange again the symbols of the issues they were focusing on at the beginning of the supervision session. The supervisees can look at their symbols in front of them and think about whether something has changed after the supervision or not. Then the group chooses the next problem they want to work on.

5. Evaluation of supervision – Objectives and helpful factors

Psychodrama psychotherapy is a complex enterprise, which is supervised on different levels. The trainee should be aware of how he or she relates to the client and how he or she establishes a working relationship. A broad foundation of methodological und technical knowledge and experience is of high significance as well. The following example of systematic reflection on supervision in psychodrama training will propose possibilities to evaluate the outcome of supervision and investigate how helpful aspects support the trainees to learn from their practice.

5.1 Focus and procedure of evaluation

Feedback from trainees is collected by two questionnaires. In a first questionnaire after each session of supervision the trainees are asked to evaluate how helpful supervision has been to them, whether they had experienced psychological distress related to the case work they are involved in, and whether this distress has been lowered or released after the supervision. All ratings of the trainees are given on a five-point Likert scale: 1=very low; 5=very strong.

Furthermore the trainees are asked to evaluate, which aspects of supervision were helpful for their psychotherapeutic work. The questionnaires are offering methodological aspects of supervision – working on a scene, a key sentence, explanation for client's psychological and social issues, biographical issues of the trainee, methodological and technical interventions –, which can be selected by the trainees as helpful for their learning.

In a second or follow up questionnaire which is answered after a period between two and four weeks later – at the beginning of the next session of supervision – the trainees are asked to evaluate their learning experiences again. This is an important step to know what can be seen as sustainable process of learning. Each trainee evaluates whether there have been changes or improvements in the relationship between the therapist and the client, the understanding of the psychological and social situation of the client and the applied techniques and methods in their therapeutic work.

5.2 Supervision setting and participating trainees

The evaluation process has been accomplished in a psychodrama training institution. The supervisor – who is also author of this chapter – is an experienced trainer who conducts the supervision groups. In this study 19 trainees (18 female, 4 male) from four different supervision groups were participating. One group consists of about 3-5 trainees. Each supervision group met within 2-4 weeks over a period of 4 months. One supervision session lasts between 4-6 working hours (1h=45 min).

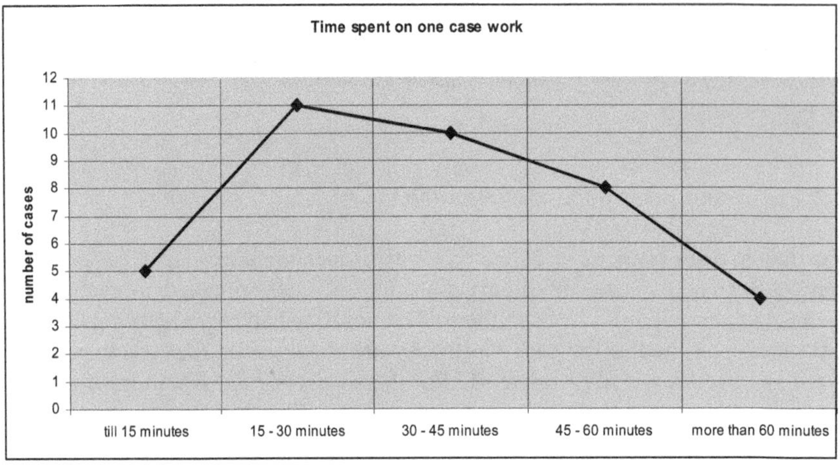

Figure 1: Time spent on one case work

In this period of 4 months 38 cases of psychotherapy were worked on (on average 3 cases per session). Most of the time supervision was done in a way which includes a broad variety of different psychodrama arrangements and techniques. The time spent on each case work varies from less than 15 minutes to more than one hour.

5.3 Results of the evaluation

In general participating trainees have a very positive attitude towards supervision. They see supervision as very helpful for their work. The results demonstrate that the majority of trainees (59,1%) say that supervision has facilitated their psychotherapeutic work on a 'strong' or 'very strong' level. The 'understanding of the psychological and social situation of the client' has increased due to their supervision. Furthermore the trainees see changes how they intervene and how they relate to the client. Last but not least the trainees report that their psychological distress, which is related to the work with the client, is reduced significantly after supervision.

5.3.1 Supervision is helpful and reduces distress in psychotherapeutic work

In general trainees characterise supervision as very helpful to them (M=4,41, SD=0,86, n=54). Trainees who present and reflect their own case work rate the help of supervision even higher (M=4,61, SD=0,73, n=36). After a period of few weeks this evaluation decreases, but remains still between 'middle' and 'strong' effects (M=3,50, SD=1,32, n=44; CI 95%, 3,1-3,9).

The figure shows that those trainees who were working on their own cases rate supervision higher than those who were reflecting on cases of others. Especially for those who were reflecting upon their own cases supervision is seen as very helpful even after some weeks (M=3,96, SD=0,93, n=23). For these trainees supervision has a sustainable effect on a high level in facilitating the psychotherapeutic work. On the other side the rating decreases from 'strong' to 'middle' effects for those trainees who were working on cases of their colleagues (M=2,79, SD=1,58, n=14).

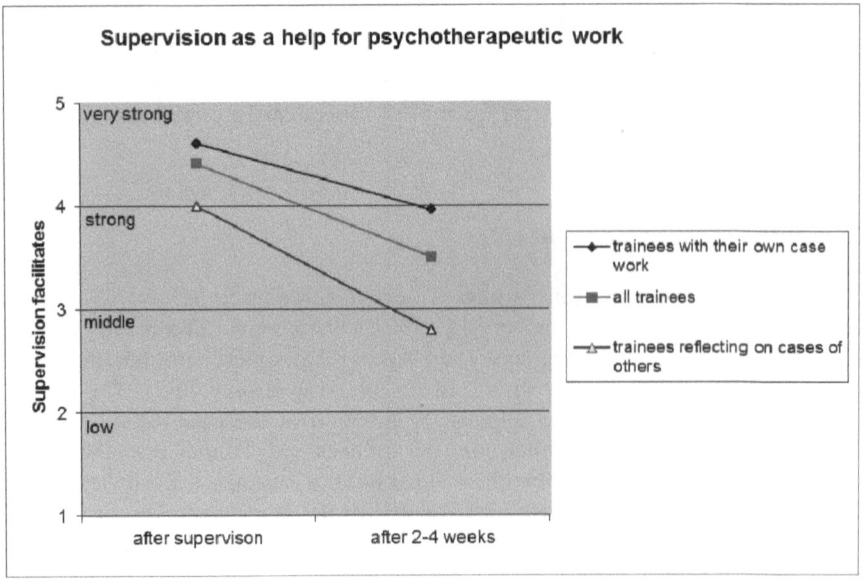

Figure 2: Supervision as a help for psychotherapeutic work

The psychological distress of the trainee which is related to their psychotherapeutic case work is before supervision 2,44 (SD=1,14, n=55) – which is between 'low' and 'middle' – and it is reduced after Supervision significantly to 1,80 (SD=0,78, n=55; t=4,87, df=54, p=.001).

5.3.2 Supervision has an impact on psychotherapeutic work of the trainee

Trainees evaluate the outcome of their supervision and respond after some weeks to the question, whether there have been changes or improvements in following aspects:

- the relationship to the client has changed;
- the understanding of the psychological and social situation of the client has improved;
- the applied techniques and methods have changed.

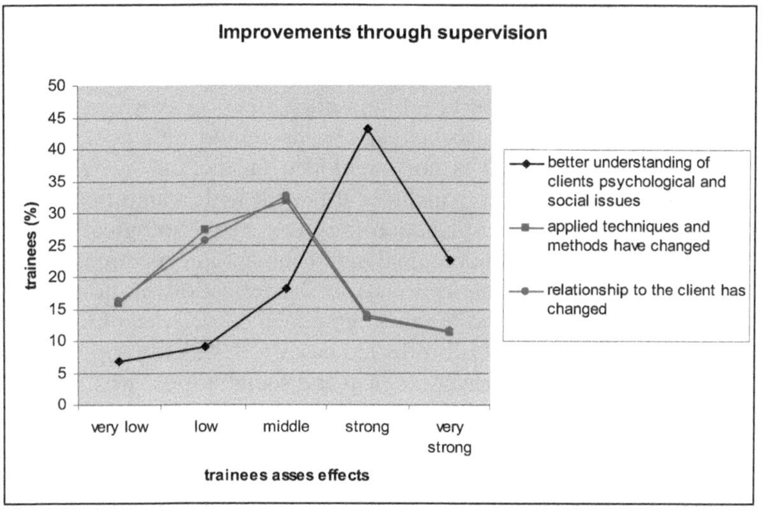

Figure 3: Improvements through supervision

	relationship to the client has changed	understanding of client`s psycho-logical and social situation	applied techniques and methods have changed
N	43	44	44
mean (M)	2,81	3,70	2,82
Standard deviation (SD)	1,20	1,07	1,23

Table 1: Results of supervision

More than two third of trainees (65,9%) confirm, that supervision has positive effects on their 'better understanding of client`s psychological and social issues' (M=3,70, SD=1,07, n=44).

In comparison to that the impact of supervision on 'applied techniques and methods' (M= 2,82, SD=1,23, n=44) or 'changed relationship to the client' (M=2,81, SD=1,20, n==43) is weaker. Both ratings show on average between 'low' and 'middle' effects.

5.3.3 Helpful methodological aspects of supervision

In order to evaluate what has been helpful in supervision the trainee could assess helpful factors right after supervision and after a period between two and four weeks. In a list of proposed aspects each trainee could either mark helpful aspects or add something what is not mentioned on the list but was helpful to them. The list of helpful aspects includes the work with a significant 'scene' on the stage, a key 'sentence', 'explanation for client's psychological and social issues', the reflection of 'methodological and technical interventions', 'emotions' (e.g. within the process of role reversal) or the reflection on how the psychotherapeutic case work is related to the biographical experience of the trainee.

In more than two third of all reflected cases (70,9%) the supervisees report that 'explanation for client`s psychological and social issues' was the most helpful aspect to them. This self-report did decrease only very little (-3,5%) after some weeks.

More than half of the supervisees report that the work with a 'scene' (58,2%) and the reflection on 'methodological and technical issues' (58,2%) was very helpful to them. Both aspects however would lose importance over time. When trainees looked back to their supervision and they rated again, the work with a significant 'scene' (-10,4%) and the verbal reflection about methodological issues (-16,9%) would not be seen so often as that important as right after supervision.

The reflection on 'emotions' (e.g. connected with role reversal) is also seen as a helpful factor in supervision (45,5%) of almost half of the trainees. Even after some weeks there is only a moderate decline of feedbacks from supervisees (-4,2%), who rate this factor as important.

One third of the supervisees report that the reflection on upcoming issues in therapeutic work which are related to his or her own biography is another helpful aspect. However, it is interesting to mention that over a short period of time this factor loses importance as well (-9,9%). The work with a key 'sentence', which is seen as an helpful factor by the trainee after supervision (32,7%) also loses importance after some weeks (-13,1%).

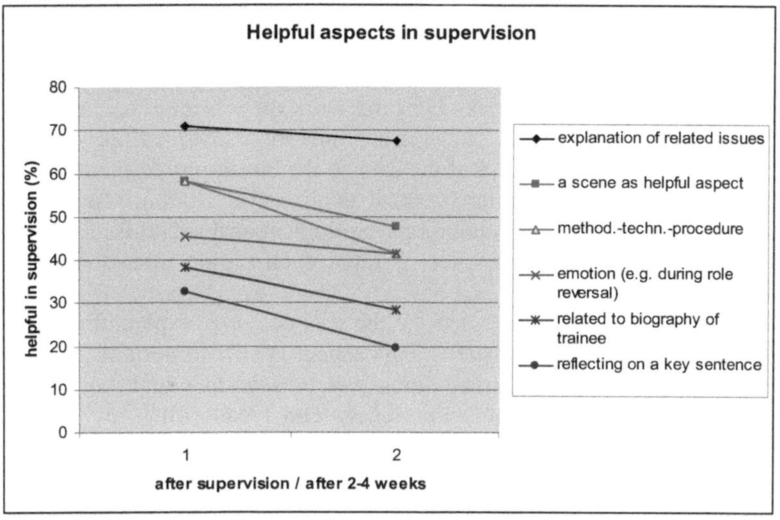

Figure 4: Helpful aspects in supervision

Helpful for psychotherapeutic work: all trainees (n=55)				trainees who reflected on their own case work (n=36)		
Helpful for psychotherapeutic work	after SV %	after 2-4 weeks%	Diff. %	after SV %	after 2-4 weeks%	Diff. %
client`s psychological/ social issues	70,9	67,4	- 3,5	69,4	78,3	+ 8,9
enacting a scene	58,2	47,8	-10,4	72,2	47,8	-23,4
method.-techn.-procedure	58,2	41,3	-16,9	58,3	39,1	-19,2
emotion (e.g. during role reversal)	45,5	41,3	- 4,2	50,0	47,8	- 2,2
related to the biography	38,2	28,3	- 9,9	38,9	21,7	-17,2
a key sentence	32,7	19,6	-13,1	44,4	30,4	-14,0

Table 2: Helpful aspects in supervision

If we look more closely and we take only the ratings of those trainees, who were reflecting on their own psychotherapeutic case work (rather than being part of the reflection about case work of somebody else in the group), we get some differences in the results. A few weeks later the work on a 'scene' seems not that much important anymore (from 72,2% it declines to 47,8%). In the same way the reflection about the relatedness of the case to the biography of the supervisee loses importance over time (from 38,9% it declines to 21,7%). On the other hand the factor 'explanation for client`s psychological and social issues', which is rated high after supervision (69,4%) is rated even higher after some weeks (78,3%).

The most helpful aspect assessed by the trainees are 'explanation for client`s psychological and social issues'. This aspect is still important when they look back after some weeks. 'Explanations' can be seen as cognitive processes which integrate different kinds of experiences and ideas which are emerging over the process of supervision.

6. Conclusion

The questions which are brought up in the supervision group are very different. Some trainees are planning a new group with certain clients; some have a question regarding theory or techniques. Sometimes ethical dilemmas within the therapeutic work need a consideration. Most of the questions of the trainees refer to the clients and the therapeutic processes. Supervision usually focuses on the questions or problems the trainee is addressing.

It is well known from research studies that the relationship between supervisor and supervisee has a strong influence on the learning of the student. However, it is not so well investigated which other aspects may have an impact on the outcome in supervision. Often students remember it was a role reversal or the mirroring technique or the reflection upon their own biographies that helped them to understand and to learn. The feedback of colleagues and supervisor which is part of the integration phase is also often reported to be helpful in supervision. However, what is seen as helpful aspect of supervision depends on what trainees remember and how these aspects are integrated in their knowledge base. Especially trainees who reflect on their own cases consider 'explanation for client's psychological and social issues' of growing importance even after some weeks, whereas all other helpful factors – one exception is 'emotion' (e.g. connected with role reversal) – are fading. We can hypothesise that the self-assessment of learning experiences is depending on how trainees are able to integrate their learning experience over time.

Supervision usually provides learning experiences in the 'here and now', but does not pay enough attention to sustainable learning. Trainees (and supervisors as well) simply can forget or 'unlearn'. The learning outcome might not be processed in a way to become part of integrated knowledge. It might be important to think about possibilities to keep 'magic moments' of reflection in supervision alive. Writing learning diaries and creating portfolios of learning processes could be considered as helpful aspects of sustainable learning in supervision.

The Sufi story in the beginning of this article should evoke our humble side as supervisors and researchers. Therefore let us keep in mind that we could get lost without the knowledge of the experienced. But still: We need more knowledge and wisdom to find our pathway through the mystery of our creative work:

Three fisher-men were sitting in a boat on a river. After some time they got thirsty and the eldest jumped out of the boat walked over the water to the beach to get some bottles of beer. An hour later the second man told the youngest to fetch again some beer for all three. The young man was not sure that he was able to do it. But his friend insisted: "Why shouldn't you be able to do it? You have seen that it is possible." The young man jumped out of the boat and disappeared in the water." The first man said after a while: Didn't you tell him where the rocks are to step on?"

References

Caspar, F. (2005). Die Entwicklung klinischer Informationsverarbeitung. In A. R. Laireiter, & U. Willutzki (Hrsg.), *Ausbildung in Verhaltenstherapie* (pp. 239-262). Göttingen: Hogrefe.

Crits-Christoph, P., Cooper, A., & Luborsky, L. (1988). The accuracy of therapists' interpretations and the outcome of dynamic psychotherapy. *Journal of Consulting and Clinical Psychology*, 56, 490-495.

Gandolfo, R. L., & Brown, R. (1987). Psychology intern ratings of actual and ideal supervision of psychotherapy. *Journal of Training and Practice in Professional Psychology*, 1, 15-28.

Gray, L. A., Ladany, N., & Walker, J. (2001). Psychotherapy Trainees' Experience of Counterproductive Events in Supervision. *Journal of Counseling Psychology*, 48 (4), 371-383.

Kahl-Popp, J. (2004). Lernziel: Kontextbezogene psychotherapeutische Kompetenz. *Forum der Psychoanalyse*. 20, 4, Springer, Medizin Verlag. 403-418.

Kellermann, P. F. (1992). Focus on psychodrama. The Therapeutic Aspects of Psychodrama. London: Jessica Kingsley Publisher.

Kraler, Ch., & Andreatta, P. (2006). *Effectiveness of Psychotherapy training – On the evaluation of psychotherapeutic competence.* SPR - Society of Psychotherapy Research: From Research to Practice, Book of Abstracts. Norderstedt, Germany.

Ladany, N., Hill, C. E, Corbett, M. M., & Nutt, E. A. (1996). Nature, extent, and importance of what psychotherapy trainees do not disclose to their supervisors. *Journal of Counseling Psychology*, 43 (1), 10-24.

Nelson, M. L., & Friedlander, M. L. (2001). A Close Look at Conflictual Supervisory Relationsships: The Trainee´s Perspective. *Journal of Counseling Psychology*, 48 (4), 384-395.

Reichelt, S., & Skjerve, J. (2002). Correspondence between supervisors and trainees in their perception of supervision events. *Journal of Clinical Psychology*, 58 (7), 759-772.

Reising, G. N. & Daniels, M. H. (1983). A study of Hogan's model of counselor development and supervision. *Journal of Counseling and Development*, 30, 235-244.

Ronnestad, M. H., & Skovholt, T. (1997). Berufliche Entwicklung und Supervision von Psychotherapeuten. *Psychotherapeut*, (5), 299-306.

Ronnestad, M. H. & Skovholt, T. M. (1993). Supervision of Beginning and Advanced Graduate Students of Counseling and Psychotherapy. *Journal of Counseling and Development*, 71, 396-405.

Ruskin, R. (1994). When Supervision May Fail: Difficulties and Impasses. In S. E. Greben, & R. Ruskin (Eds.): *Clinical Perspectives on Psychotherapy Supervision* (pp. 213-261). Washington: American Psychiatric Press.

Williams, A. (1995). *Visual and Active supervision.* New York, London: W.W. Norton & Company.

Worthen, V., & McNeill, B. W. (1996). A Phenomenological Investigation of "Good" Supervision Events. *Journal of Counseling Psychology*, 43 (1), 25-34.

Contributors

Apter Norbert, M.Ed., a graduate of Harvard; trainer, supervisor, psychodramatist and psychotherapist specialised in both the person-centred approach (Carl Rogers) and psychodrama (J.L. Moreno), private practice in Geneva. *Key activities:* Individual and group work in numerous countries, e.g. teambuildings, and conflict resolutions within teams, and a great number of training courses for private practicioners or professionals in institutions, in companies and in international organisations.

Artzi Einya, coordinator of the department of psychodrama of the expressive therapy training centre at the Kibbutzim College of Education; co-founder & key member of the Israeli Psychodrama Association. *Key activities:* Psychodramatic work with trauma and post-trauma in Jewish and Arab war survivor populations, director and facilitator of psychodrama workshops at universities, colleges and centres for mental health throughout Israel; author of the book *Psychodrama*, Dvir Publishing House, Tel Aviv, 1991.

Boria Giovanni, Ph.D., director of the School of Specialization *Studio di Psicodramma* in Milan; clinical psychologist, psychotherapist, psychodramatist; founder of the *Studio di Psicodramma*, first centre in Italy of experimentation and diffusion of Morenian methodology; founding member of the A.I.Psi.M (Italian Association of Morenian Psychodramatists) and FEPTO, first president of FEPTO; published many articles and books on classical psychodrama.

Chesner Anna, co-director of the London Centre for Psychodrama; psychodrama psychotherapist and supervisor at the Maudsley Hospital and in private practice in London, member of the Dramatherapy MA team at Anglia Ruskin University, Cambridge; founding member of London Playback Theatre; several publications on supervision and psychodrama techniques. *e-mail:* chesnera@aol.com,
web: www.londoncentreforpsychodrama.org

Daniel Sue, director of the Psychodrama Institute of Melbourne; psychodrama trainer, educator and practitioner, consulting psychologist and psychotherapist in private practice in Melbourne, Australia; founder of the Moreno Psychodrama Society; member of the Board of IAGP and chairperson of the Australian and Aotearoa (New Zealand) Board of Psychodrama. *Key activities:* Conducting workshops and seminars worldwide, writing, researching and teaching role theory. *e-mail:* suedan@netspace.net.au,
web: http://www.psychodrama-institute-melbourne.com

Dudler Agnes, Dipl.-Psych., psychologist and psychotherapist in private practice in Bonn; trained in client centred therapy, breath therapy, bioenergetics, yoga, systemic, gestalt, psychoanalytic therapy and meditation; founder of the "Institut für Psychodrama

Szenen"; member of the DFP/DAGG. *Key activities:* Counsellor, psychotherapist with individuals, couples and groups, supervisor and coach, sociodrama with large groups, training in awareness and self care; published several articles about psychodrama with individuals, about the "role atom" and sociodrama. *e-mail:* agnes@dudler.org

Erdélyi Ildikó, Prof. em. Dr., Institute of Psychology of Károli G. R. University (Budapest); clinical psychologist and trainer, psychotherapist for psychodrama and supervisor in private practice, training psychoanalyst at the Society for Psychodrama and the Society for Psychoanalysis; first vice-president of the Hungarian Society for Psychodrama. *Key activities:* Research on the effect mechanism of psychodrama (protagonist-centred and group-centred) psychotherapy; several publications on psychodrama related themes.
e-mail: dr.erdelyi.ildiko@gmail.com, *web:* http://www.erdelyiildiko.hu

Fürst Jutta, Dr. phil., scientific chair of a training programme for psychodrama psychotherapy at the University of Innsbruck; clinical psychologist, supervisor and psychotherapist for psychodrama and guided affective imagery, in private practice; past president of FEPTO. *Key activities:* Trainer and lecturer for psychodrama at several universities and of the Austrian Society of Groupdynamics and Grouptherapy; workshop leader in various European countries; author of articles and co-editor of a book on psychodrama therapy.
e-mail: jutta.fuerst@uibk.ac.at, *web*: http://members.cnh.at/jutta.fuerst

Fontaine Pierre, MD, Prof. em. of Child & Family Psychiatry, Faculty of Psychology, University of Louvain; clinical psychologist; former head of clinical department of Child Psychiatry Louvain; founder and former director of the psychodrama training school CFIP-Verveine Brussels, past member of the Board of Directors of IAGP; former vice-president of FEPTO. Last *key activities:* collaborative research with poor persons; editor of *Psychodrama Training. A European View* (2nd ed. 2001: FEPTO Publ.).
e-mail: p.fontaine@uclouvain.be

Gasseau Maurizio, Ass. Prof. of Dynamic Psychology at the Università della Valle D'Aosta; Jungian analyst and psychodrama trainer; former vice president of FEPTO, currently chair of the psychodrama section of IAGP; co-developed Jungian Psychodrama; *Key activities:* Researching dreams in psychodrama; lecturer on Jungian psychodrama, dreams and psychodramatic social dreaming matrix at several universities and author of 90 publications; book: *From Analytical Psychology to Jungian Psychodrama* 2009, Franco Angeli. *email:* m.gasseau@univda.it, *web:* www.jungianpsychodrama.com

Gött Hilde, Dipl.Soz.päd., psychotherapist for children and adolescents, therapist for addicted, counselor and supervisor, trainer and supervisor for psychodrama; chairwoman of the Psychodrama Institute for Europe (PIfE). *Key activities:* Supervision and workshops in various European countries, working in different NGOs focused on trauma, domestic violence, sexual abuse and suicide, working on transmission of trauma during the Holocaust for offspring of the victims and perpetrators; various articles on this topic.
e-mail: hildegoett@t-online.de, *web:* www.pife-europe.eu

Karapostoli Natassa, occupational therapist, sociotherapist, psychodramatist, family therapist partly in private practice; secretary of the Training and Research Department and member of the board of the Open Psychotherapy Centre, member and board member of several societies, (i.e. National Organization of Psychotherapy of Greece, Association of Therapeutic Community, Psychodrama and Sociotherapy etc.).

Kayır Arşaluys, Dr. M.Sc. Univ. Prof. em. of clinical psychology and faculty member of the Istanbul University Medical Faculty Department of Psychiatry; trainer and supervisor at Dr. Abdulkadir Özbek Psychodrama Institute and Sexual Education, Treatment and Research Association, FEPTO advisory board member. *Key activities:* Psychodramatist, sexual therapist, specialised on treating sexual dysfunctions in group work using psychodrama; main publications about sexuality, psychodrama and group processes of women with vaginismus. *e-mail:* arkayir@gmail.com

Krall Hannes, ao. Univ.-Prof. Dr., University of Klagenfurt, Educational Sciences and Research; pedagogue and psychologist, counsellor, psychotherapist, supervisor, trainer at the Austrian Society of Groupdynamics and Grouptherapy and lecturer at the University of Innsbruck; chair of the FEPTO Research Committee. *Key activities:* Current research interest in psychodrama training, supervision and psychodrama practice. Several publications about psychodrama, supervision, violence, trauma of children and youth.
e-mail: hannes.krall@aau.at

Lap Jan, Practical Theologist, University of Tilburg (The Netherlands); educated in pastoral therapy, supervision, group dynamics and consultancy for organisations, director of psychodrama, general manager, trainer and supervisor at the Institute LapStreur SE, Education and Development of Psychodrama & Bibliodrama; board member NBES (Dutch-Belgian Examination Foundation for CP & TEP). *Key activities:* Specialised in psychodramatic bibliodrama, working with concise psychodramatic action forms.
e-mail: info@lapstreur.nl, *web:* www.lapstreur.nl

Moita Gabriela, Ph.D. in Biomedical Sciences, lecturer at the Superior Institute of Social Work of Porto; clinical psychologist, psychodramatist, psychodrama trainer and supervisor; co-chair of FEPTO, former chair of FEPTO Research Committee, chair of the SPP (Portuguese Psychodrama Association), Vice-chair of SPSC (Portuguese Society of Clinical Sexology). *Key activities:* Work with psychodrama in psychotherapeutic and educational field. *e-mail:* gabriela.moita@mail.telepac.pt

Nève-Hanquet Chantal, psychologist, psychodramatist at CFIP (Belgium); Jungian analyst at the SBPA, trainer in family therapy, trainer for psychodrama; treasurer of FEPTO. *Key activities:* Psychotherapy with individuals, couples and families in private practice, supervision groups and workshops in various European countries; published many articles about the landscape genogram with Jacques Pluymaekers and about the 'body in therapy' with Christine vander Borght. Co-author of several articles with Pierre Fontaine. *e-mail:* hanquetchantal@gmail.com

Perrotta Leandra, teacher at the Università della Valle D'Aosta and the International Academy of Psychogenealogy in Bologna; clinical psychologist, psychodrama trainer and dancetherapist; board member of FEPTO and the Associazione Mediterranea di Psicodramma. *Key activities:* Developed an integrative model of Jungian psychodrama and dancetherapy, transgenerational work, workshop leader in various countries all over the world; published many articles on dream work and Jungian Psychodrama.

Teszáry Judith, supervisor and educator in private practice and at Stockholm City Social Administration, psychodrama trainer and supervisor; founding member and former president of FEPTO, member of the Scientific Program Committee of IAGP, chair of the Swedish Association of Psychodramatists, founder & editor of the Swedish Psychodrama Journal. *Key activities:* Senior trainer and supervisor in various countries, psychodrama and sociodrama in therapeutic settings, foster care assessment.
e-mail: judith.teszary@comhem.se

Roma-Torres António, MD, M.Sc. in Psychiatry and Mental Health Professional development, head of Psychiatry Department in São João Hospital Center; family therapist; founding Member of FEPTO and former chair of FEPTO Research Committee, former board member of IAGP, former chair of the SPP (Portuguese Psychodrama Association. *Key activities:* Research and clinical practice on eating disorders, psychiatrist in public and private practice. *e-mail:* aroma@sapo.pt

Tsegos Ioannis K., psychiatrist and group analyst (I.G.A. London), director of the Training and Research Department of the Open Psychotherapy Centre and psychotherapist in private practice; president of Institute of Group Analysis (I.G.A.) Athens and Group Analytic Society (G.A.S.) Greece, founder of the European Group-Analytic Training Institutions Network (E.G.A.T.I.N.). *Key activities:* Director of the book publication series in Greek entitled "Contemporary Psychotherapy".

Weiß Kersti, Dipl.-Psych., tutor for supervision (DGSv) and director of the department of supervision at the Institute for Organizational Development and Supervision; certified psychotherapist, psychodramatherapist (DFP/DAGG), supervisor and coach. *Key activities:* Coaching, organisational development and advanced training, trainer for supervision, psychodrama and organizational development; main publications on guidance, conflicts, team building etc.
e-mail: info@kersti-weiss.de, *web:* www.kersti-weiss.de, www.ipos-ekhn.de

Kolibri 360 GmbH
Ferdinand-Jühlke-Straße 7
99095 Erfurt, Deutschland
produktsicherheit@kolibri360.de